Evolutionary Genetics and Environmental Stress

Evolutionary Genetics and Environmental Stress

ARY A. HOFFMANN

Department of Genetics and Human Variation
La Trobe University
Bundoora, Victoria
Australia

and

PETER A. PARSONS

Department of Zoology
University of Adelaide
Adelaide, South Australia
Australia

Oxford   New York   Tokyo
OXFORD UNIVERSITY PRESS
1991

Oxford University Press, Walton Street, Oxford OX2 6DP
Oxford New York Toronto
Delhi Bombay Calcutta Madras Karachi
Petaling Jaya Singapore Hong Kong Tokyo
Nairobi Dar es Salaam Cape Town
Melbourne Auckland
and associated companies in
Berlin Ibadan

Oxford is a trade mark of Oxford University Press

Published in the United States
by Oxford University Press, New York

British Library Cataloguing in Publication Data
Hoffmann, Ary A.
Evolutionary genetics and environmental stress.
1. Organisms. Evolution
I. Title II. Parsons, Peter A.
575
ISBN 0–19–857732–X

Library of Congress Cataloging in Publication Data
Hoffmann, Ary A.
Evolutionary genetics and environmental stress/Ary A. Hoffmann
and Peter A. Parsons.
Includes bibliographical references and indexes.
1. Variation (Biology)—Environmental aspects. 2. Genetics.
3. Evolution. 4. Chemical evolution. I. Parsons, P. A.
(Peter Angas), 1933– . II. Title.
QH401.H64 1991 575.1'3—dc20 90–40177
ISBN 0–19–857732–X

Typeset by Cotswold Typesetting Ltd, Cheltenham
Printed in Great Britain by
Courier International Ltd, Tiptree, Essex

Preface

The interaction between organism and environment is central for understanding evolution. Unless organisms can continuously change to accommodate the constantly altering physical and biotic environments to which they are exposed extinctions will follow, as documented throughout the fossil record. From the molecular to the biogeographic levels of organization, there is accumulating evidence indicating that extreme and potentially injurious environmental conditions underlie much evolutionary change. Even though periods of such severe stress are unpredictable in occurrence and of short duration, fundamental changes in the natural order of systems are likely at these times, both in terms of species extinctions and bursts of evolutionary change within species. In contrast, the literature of evolutionary biology has tended to focus upon severe stress rather infrequently; this applies particularly to laboratory studies which are normally carried out under conditions that are experimentally easier and tend to be optimal.

The emphasis of this book is therefore on genetic changes in populations at the extreme end of the stress gradient at the limits of resistance. We demonstrate that in some situations severe stress can be regarded as an ' environmental probe', leading to generalizations which are difficult to perceive under less extreme conditions. For example, biological systems under stress can be described in terms of energetic costs, from which suggestive associations between habitat, life-history characteristics, and stress resistance emerge. This book is necessarily multidisciplinary in approach, with an emphasis upon the interface of ecology, genetics, physiology, and the study of behaviour and development. This follows from the integrative nature of any attempt to understand evolutionary processes. In biology today, we are facing the dilemma that the ability of molecular biologists to describe the genome of an organism in molecular terms is far outstripping the application of this knowledge to the organismic and higher levels of organization. Consequently, it is difficult to handle the interaction of many disciplines in an era of ever increasing specialization. Our task is therefore daunting, but an emphasis on stress permits some simplifications and generalizations to be made. The usefulness of our approach will be shown in the last chapter in terms of a discussion of the range expansion of species, and of conservation strategies under rapidly changing conditions.

v

Books are not normally written in isolation. We are most grateful to Y. Fripp, J. Hughes, R.K. Koehn, S. McKechnie, M. Turelli, and P. Ward for reading and commenting perceptively on selected chapters. During the writing process, we have been assisted by discussions with numerous people, in particular R.W. Allard, J.H. Bennett, F. Cohan, B. Geer, P. Grime, L. Harshman, M. Kohane, E. Mayr, and P. Ward. Excellent secretarial help was provided by L. Gigliotti, F. Pizzey, and M. Tarzia at La Trobe University, and Louise Parsons provided assistance with indexing. Finally, work on this book for one of us (P.A.P) commenced while Visiting Professor at Griffith University; grateful thanks are given to the Vice-Chancellor, Professor L.R. Webb, and especially to Professor R.S. Holmes, Dean of Science and Technology, for helping to make this period so rewarding.

Bundoora and Adelaide   A.A.H.
July 1990   P.A.P.

Contents

1. Introduction

1.1 Stress definitions

We commence with a definition of 'stress' to be used in this book because this term can take a number of meanings in a wide range of contexts. Discussions of stress in biology recognize two components, external or internal forces that are applied to organisms or other biological systems, and changes in biological systems that occur as a consequence of these forces. These two components are usually considered together because the impact of a force cannot be defined without reference to a biological system, and the term 'stress' has been applied to the change in the biological system, the force or both components simultaneously.

A definition of stress based on changes in biological systems was emphasized by Hans Selye, who recognized stress by its physiological effects on mammals, defining it to be a state 'manifested by a specific syndrome which consists of all the non-specifically induced changes within a biological system' (Selye 1956). Selye referred to forces responsible for these changes as 'stressors'. He suggested that, depending upon the length and severity of the stress, an animal passes through the following three phases:

(1) alarm when the organism is not capable of resisting the stress;

(2) resistance when the body returns to normal functioning;

(3) exhaustion when the resistance is lost.

These three phases represent the sum of the non-specific reactions of the body resulting from long exposure to chronic stress. The reactions include many histological, morphological, biochemical, and functional alterations, and a variety of biotic and physical factors result in the same alterations (Munday 1961). The reactions are mediated by hormones, starting with excitation of the hypothalamus by the stressor and followed by upsets in the adrenal/pituitary system. Selye defines stressors widely to include any factor that induces a non-specific reaction, which comprise not only external physical and chemical factors but also such activities as exercise or courtship. While Selye's work was restricted to mammals and is medically orientated, it emphasizes the concept of various stressors having similar consequences at the physiological level.

Some biological definitions of stress have emphasized more drastic changes than those outlined by Selye, but maintained the emphasis on the biological system rather than on a force. For example, Bayne (1975) defined stress as:

a measurable alteration of a physiological (or behavioural, biochemical or cytological) steady state which is induced by an environmental change, and which renders the individual (or the population, or the community) more vulnerable to further environmental change.

Stress in this definition is the disturbance of the normal functioning of a biological system by any environmental factor, as detected by departures from a steady state.

Other biologists have applied stress to environmental forces rather than an organism's state. For example in plants, Grime (1979) defined stress as 'environmental constraints, shortages and excesses in the supply of solar energy, water and mineral nutrients' and 'sub- or supra-optimal temperatures and growth inhibiting toxins'. These are basically external constraints on dry matter production that disturb normal functioning. With reference to animals, Sibly and Calow (1989) defined stress as 'an environmental condition that, when first applied, impairs Darwinian fitness', including environmental conditions that decrease survivorship and fecundity as well as growth. Similarly, Koehn and Bayne (1989) refer to stress as 'any environmental change that acts to reduce the fitness of an organism'. Both these definitions emphasize environmental forces, and also recognize that these forces can be identified by their effects on biological systems. Levitt (1980) and others have also argued for the necessity to incorporate the environmental force and biological system into a definition of stress. Levitt points to the mechanical definition of stress as a situation where a force is placed on a body which results in a state of strain on that body. This definition emphasizes the integrated nature of the two components mentioned above because the deleterious effects of an environmental factor cannot be recognized without reference to a biological system.

The discussions of Grime, Levitt, Bayne, and others focus on a much narrower range of forces than Selye's broad approach, specifically emphasizing physical factors of the environment rather than parasites, diseases, or behavioural stress from factors such as crowding or confinement. The physical factors include climatic variables such as temperature and humidity, as well as radiation, food shortage, pollutants, pesticides and other environmental toxins. The effects of these physical factors on organisms may depend on their interactions, because exposure to extremes of one factor often makes organisms more susceptible to other factors. This has been emphasized in much recent literature on environmental pollution, but was appreciated by Darwin (1859) with respect to the interaction between climatic factors and other factors as indicated by the following:

Even when climate, for instance extreme cold, acts directly, it will be the least vigorous individuals, or those which have got least food through the advancing winter, which will suffer most.

While physical features of the environment are considered as the primary stressors, they may interact with biotic factors such as competition, predation, and parasitism to cause changes in biological systems. Damage from physical extremes of the environment may make organisms more susceptible to disease and parasitism. The effects of parasites and diseases are often most evident in organisms that are exposed to sub-optimal conditions, and diseases have in fact been used to monitor pollution effects in organisms such as fish (Wedenmeyer et al. 1984). Competition may have a more indirect effect by altering the physical environment. For example, competition between two plants for moisture, light, or soil nutrients could result in altered physical conditions such that these factors become limiting for growth and reproduction. Competition and predation may also exacerbate the effects of physical factors. Organisms that are weakened by exposure to climatic extremes will be more prone to predation and less competitive.

The definitions of stress used by Grime (1979) and Sibly and Calow (1989) encompass a broad physical range of conditions. For example, conditions causing a slight reduction in plant yield from an optimal level would be considered stressful—and indeed it is in this sense that the term 'stress' is often used in the agricultural literature. In the wild this would mean that all organisms are stressed most of the time, because survival and reproduction probably never achieve their maximum under natural conditions. Nevertheless, the term 'stress' tends to be applied in the evolutionary and ecological literature to situations where there is high mortality (or the potential for high mortality) or a drastic reduction in reproductive output because of changed environmental conditions. Implicit in these applications is the idea that the persistence of stressful conditions will result in permanent damage because these conditions are near or beyond the limits of resistance (Levitt 1980; Blackstock 1984). These situations are of evolutionary interest rather than small departures from optimal conditions because they indicate periods of intense selection and rapid evolutionary change as well as the potential for extinction.

As emphasized by Odum et al. (1979), stress can be viewed as an agent placing an organism or an ecosystem at a disadvantage, since it requires a continued expenditure of excess energy which is incompatible with survival. However, environmental forces at low levels may occasionally lead to a favourable deflection or a 'subsidy' at the other end of the continuum. Stress is a detrimental or disorganizing influence, while subsidy is an input producing a beneficial response, even though it may be accompanied or followed by negative responses. Odum *et al.* (1979) cite moderate flooding which would tend to enhance productivity while high level flooding reduces it. This continuum is illustrated by the subsidy–stress gradient in Fig. 1.1. The gradient is relevant to other stressors such as toxic substances that are typically regarded as totally deleterious, but may be advantageous at low levels, a phenomenon referred to as hormesis (Calabrese *et al.* 1987). Hormesis has been observed in plants and animals for many classes of chemicals, and implies a higher fitness in the presence of a chemical agent at low concentrations than at zero concentration (Parsons 1989*a*). At higher concentrations, fitness falls progressively as the agent becomes increasingly stressful.

In this book, we will follow Levitt (1980) and others by using the term 'stress' to represent an environmental factor causing a change in a biological system which is potentially injurious. The altered state of the

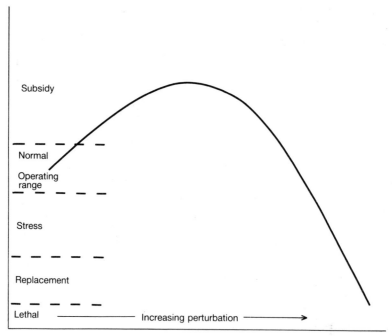

Fig. 1.1. Hypothetical performance curve for a perturbed ecosystem, indicating the subsidy–stress gradient. (Simplified from Odum *et al.* 1979.)

biological system will be referred to as 'stressed' or 'under stress', while environmental factors responsible for the stressed state will be referred to as 'environmental stresses', 'stressors', or 'stressful conditions'. Because our emphasis is on genetic aspects, the biological system will usually refer to an organism rather than a community or a cellular level. Environmental stresses will refer to physical rather than biotic factors, in agreement with most discussions on stress in biology. Biotic factors will be considered only to the extent that they interact with physical factors. The emphasis will be on extreme environmental conditions that are potentially injurious. The energetic costs and departures from homoeostasis that can be used to describe biological systems under stress will form an important component of our discussion.

1.2 Environmental stress and genetics: from Darwin to the present

In 1859, Darwin published *On the origin of species by means of natural selection*, in which he considered how one species evolves into a different species in terms of individual organisms and populations. Darwin put forward the view that in any environment an organism will accumulate over time the variations best fitting it to its surroundings. As the environment changes, new variants will become advantageous, and will tend to supplant variants that have become less well adapted. He appreciated that the effectiveness of this process of natural selection depends upon the variants being inherited and available at times of environmental change.

Darwin (1859) greatly emphasized the significance of competition in the process of natural selection and hence in evolutionary change:

As the species of the same genus usually have, though by no means invariably, much similarity in habits and constitution, and always in structure, the struggle will generally be more severe between them, if they come into competition with each other, than between species of distinct genera.

As noted by Mayr (1963), Darwin tended to refer to competition in more dramatic terms than is usually assumed today. In some cases, his discussions of competition invoke physical features of the environment:

Hence, as more individuals are produced than can possibly survive, there must in every case be a struggle for existence, either one individual with another of the same species, or with individuals of a different species, or with the physical conditions of life.

Eventually under conditions of extreme stress, he writes that:

Not until we reach the extreme confines of life, in the Arctic regions or on the borders of an utter desert, will competition cease. The land may be extremely cold or dry, yet there will be competition between some few species, or between the individuals of the same species, for the warmest or dampest spots.

Darwin therefore envisaged a continuum ranging from competition being exceedingly important in benign environments, to being of minor importance in extreme environments where the effects of environmental stresses predominate.

While not discussed in the same depth as competition, Darwin refers directly to extreme stress on a number of occasions:

Climate plays an important role in determining the average numbers of a species, and periodical seasons of extreme cold or drought seem to be the most effective of all checks.

Hence in going northwards, or in ascending a mountain, we far oftener meet with stunted forms, due to the directly injurious action of climate, than we do in proceeding southwards or descending a mountain. When we reach the Arctic regions, or snow-capped summits, or absolute deserts, the struggle for life is almost exclusively with the elements.

But a plant on the edge of a desert, is said to struggle for life against the drought.

and in Darwin and Wallace (1859), Darwin wrote:

It should be remembered, that in most cases the checks are recurrent yearly in a small, regular degree, and in an extreme degree during unusually cold, hot, dry, or wet years, according to the contribution of the being in question.

Even so, Darwin regarded evolution to be predominantly a gradual process emphasizing intricate and complex interactions among species, and to a lesser extent, between species and their physical environment.

This last quotation comes from the joint publication of the theory of evolution by natural selection by Darwin and A.R. Wallace in 1858. As Mayr (1982) points out, Wallace based his conclusion on a rather strictly ecological argument, and avoided the debate concerning the morphological status of species and varieties. He concluded that the population size of a species is determined by natural checks on potential population increases. Wallace considered that an enormous number of animals must die each year to keep the population size constant. Those that die are the weakest, leaving the most healthy and vigorous who can obtain food and avoid predation. Wallace's major work following the publication on the theory of evolution was his *Geographical distribution of animals* (1876), the classic of zoogeography for many decades, and where he wrote:

Climate appears to limit the range of many animals, though there is some reason to believe that in many cases it is not the climate itself so much as the change of vegetation consequent upon climate that produces the effect.

Any slight change, therefore, of physical geography or of climate, which allows allied species hitherto inhabiting distinct areas to come into contact, will often lead to the extermination of one of them; and this extermination will be effected by no external force, by no actual enemy, but merely because the one is slightly better adapted to live, to increase, and to maintain itself under adverse circumstances, than the other.

His emphasis upon climate in determining the distribution of species is clear, but more in terms of the mean than the extremes to which Darwin tended to refer. Perhaps this is because Wallace's major periods of field work were in the tropics. Indeed in 1878 he wrote:

It is difficult for an inhabitant of our temperate land to realize either the sudden and violent contrast of the arctic seasons or the wonderful uniformity of the equatorial climate.

There are few discussions of competition by Wallace. However in the *Geographical distribution of animals*, quoting Darwin whom he respected greatly, he discusses complex relations between animals and plants in a region, and conceded that:

The range of any species or group in such a region, will in many cases (perhaps in most) be determined, not by physical barriers, but by the competition of other organisms.

Even so, Wallace mainly emphasized physical variables as major determinants in geographical distribution and evolutionary change.

Environmental stress was considered rather little in the development of the genetic theories of evolution by R.A. Fisher, J.B.S. Haldane and Sewall Wright. In his discussion of adaptation, defined as the degree of conformity between organism and environment, Fisher (1930) wrote:

An organism is regarded as adapted to a particular situation, or to the totality of situations which constitute its environment, only in so far as we can imagine an assemblage of slightly different situations, or environments, to which the animal would on the whole be less adapted; and equally in so far as we can imagine an assemblage of slightly different organic forms, which would be less well adapted to that environment.

Because the environments of the living world have unpredictable components over time and space, perfect adaptation is not possible. Fisher (1930) developed a model of adaptation of existing organisms in relation to environmental and organismic change by comparison with a hypothetical perfectly adapted organism. Very small, undirected changes in either organism or environment approach a 50 per cent chance of being advantageous (Fig. 1.2) while very great changes are always maladaptive; even if in a favoured direction, organisms would end up in less than optimal environmental situations by overshooting points of

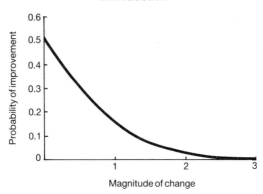

Fig. 1.2. The relation between the magnitude of an undirected change and the probability of improving adaptation when the number of dimensions of the environment is large. (Adapted from Fisher 1930.)

increased adaptation. Thus, as the magnitude of change increases and irrespective of the number of 'dimensions' used to assess the environment, the probability of improvement in adaptation will decrease towards zero. By parallel arguments, at the genetic level, new mutations will be expected to have a greater probability of having deleterious effects if they have large phenotypic effects, emphasizing the importance of genes of small effect (micromutations) in adaptation to an environment. The existence of genetic variation necessarily means that perfect adaptation is an unachievable aim, so that at the level of gene, organism, and environment, perfect adaptation must be regarded as a theoretically unattainable goal, especially during major environmental shifts.

Although this scheme can be interpreted in terms of adaptation to conditions representing varying stress levels, Fisher (1930) considered the physical factors of the environment to be less important than interactions among organisms, as indicated by the following:

Probably more important than the changes in climate will be the evolutionary changes in progress in associated organisms. As each organism increases in fitness, so will its enemies and competitors increase in fitness; and this will have the same effect, perhaps in a much more important degree, in impairing the environment, from the point of view of each organism concerned.

Haldane (1932) also focused on competition in *The causes of evolution*, although he noted the importance of extreme periods of stress in the context of characters which appear useless but have been selected under rare extreme conditions. He described the example of sponges that live between tide-marks and have elaborate systems for ejecting water, and argued that such systems may have been selected in situations where rainstorms or excess heat kill most of the population but leave a few survivors in stagnant water where ejection systems may be important.

Following this phase was the appearance of increasing numbers of books from about 1936 to 1947 establishing and developing the synthetic theory of evolution (Mayr 1982). A particularly important landmark was Schmalhausen's (1949) book, *Factors of evolution*, which first appeared in the USSR in 1947. This book has substantial discussions of genotype by environment interactions including a consideration of extreme environments. Arguing from laboratory selection experiments for cold resistance in *Drosophila*, Schmalhausen appreciated that natural selection for resistance to environmental extremes could be effective over quite short time periods. Even so, he considered that all really new reactions of the organism are never adaptive, which follows from Fisher's (1930) analysis. He proposed that every genotype is characterized by its own specific 'norm of reaction', which includes adaptive modifications to different environments. Limits to the 'norm' will be established by criteria such as reduced survival and reduced reproductive rates, and will depend upon the types of conditions encountered by organisms in nature. The potential for interactions between the genotype and the environment within these limits is very substantial as will be discussed in Chapter 7.

Schmalhausen's view was that 'progressive evolution consists of a continuous acquisition of new reaction norms', and it can be surmised that he regarded this most likely under substantial environmental perturbations. He emphasized the importance of climatic stress in mountainous, continental and polar regions and provided many examples of differences within species in sensitivity to diverse temperatures in *Drosophila*, various fishes, rodents, lizards and plants. Schmalhausen cited data of Timofeef-Ressovsky on viabilities at various temperatures in *Drosophila funebris*, which indicated that differences between populations are in accord with the climatic features of habitats where they originated. In particular, the continental populations studied by Timofeef-Ressovsky (1940) were highly resistant to high and low extremes of temperature.

Many other early studies of physiological races which are reviewed in Mayr (1963) demonstrated that populations had diverged genetically in response to physical features of the environment, particularly to temperature. These studies include the genecological approach developed for plant populations by Turesson (1922), Clausen, and others. In this approach plants of the same species from climatically diverse locations are reciprocally transplanted or grown in a common garden to demonstrate that major differences between them tend to be genetic. For example, Clausen *et al.* (1940) studied races of a member of the rose family, the semi-perennial plant, *Potentilla glandulosa*, by taking individuals from sea-level, mid-altitude and alpine zones of the Sierra Nevada region of California, and then transplanting across climatic habitats in all possible combinations. The transplants often died or grew poorly in foreign habitats, while performing relatively better in their native habitats, indicating the existence of climatic races.

Early genetic studies also demonstrated genetic variation for resistance to environmental stresses within populations. In particular, several experiments related karyotypic variation in *D. pseudoobscura* to resistance to physical factors. For example, Heuts (1948) found that relative humidity influenced the pupal eclosion of karyotypes at 25 °C, while Spassky (1951) and Levine (1952) found that the viability of karyotypes was differentially affected by combinations of humidity and temperature. More sophisticated genetic experiments were carried out on stress resistance in *Drosophila melanogaster*, where genetic differences between strains were located to the chromosome level for resistance to insecticides (Crow 1957) and other chemical stresses, as well as radiation resistance (reviewed in Parsons 1974).

An attempt to link genetic variation and environmental stress with development was made by Waddington (1953, 1956). He described situations where sub-lethal environmental perturbations, including extreme temperatures, reduced the stability of development when applied at critical stages and generated phenotypic abnormalities. These abnormalities were often so great that survival in nature would be unlikely. The association between developmental processes and stress has also been considered in much of the early research on fluctuating asymmetry, which is a measure of the difference between bilateral characters. Examples involving temperature and chemical stresses include Thoday (1958) and Parsons (1961, 1962). The degree of asymmetry changes as stress levels increase and these changes are thought to represent underlying developmental instabilities, as will be discussed further in Chapter 5.

In recent years, there has been increasing emphasis on genetic variation at the protein and DNA levels detected by techniques such as electrophoresis. This has led to a debate as to whether or not the molecular variation is affected by natural selection. Discriminating between these possibilities has been difficult because of the problem of extrapolating from the molecular to the organismic levels. Even so, gene frequencies detected by electrophoresis often tend to be correlated with temperature and other variables at the geographic level, and it has been possible to relate variation at some loci to environmental stresses. In fact, much of the recent emphasis on demonstrating selection at single loci has been on characterizing genotypic differences under extreme conditions (e.g. Watt 1985; Hartl *et al.* 1985; Nevo 1988).

Molecular studies have also provided insights into the types of genes that are involved in responses to environmental stress. In particular, there is one class of proteins that appears to be closely associated with a variety of stresses, the so-called heat-shock proteins or stress-induced proteins. These proteins may be important in the acquisition of increased resistance to high temperatures and other stresses during acclimation in a range of species (Lindquist 1986).

In contrast to the molecular approach, ecologists have emphasized quantitative traits of significance in determining the distribution and abundance of organisms, based on the assumption that the organism is the unit of selection. An aim of these studies is to dissect fitness into its life-history components and to investigate selection on these components in natural populations. There has been controversy as to which traits should be studied in detail. In their classic book, *The distribution and abundance of animals,* Andrewartha and Birch (1954) emphasized the importance of physical features of the environment, while competition and other interactions between organisms were emphasized by Lack (1966) and others. Andrewartha and Birch studied insect populations, which often fluctuate in response to physical factors, while Lack worked with birds, which often show more constant population sizes, so these divergent views partly reflect the groups of organisms that were studied. This debate continues today (e.g. Salt 1984), and has resulted in an emphasis on survival under climatic extremes on the one hand, and traits associated with competitive ability on the other hand. These different emphases have carried over to genetic studies of quantitative variation. For example, in *Drosophila* there were numerous experiments that considered variation in competitive ability (reviewed in Barker 1983), while many other genetic experiments have examined variation in stress resistance traits (Parsons 1974). Similarly, reviews on evolutionary ecology in *Drosophila* range from those that almost completely ignore environmental stress and focus almost entirely on competition (e.g. Mueller 1985) to those where environmental stress forms the main emphasis (e.g. Parsons 1983).

A recent development in evolutionary genetics has been the re-emergence of quantitative genetic studies of natural populations. This has been stimulated partly by theoretical developments on the evolution and analysis of quantitative traits (Barton and Turelli 1990), and partly by a growing interest in life-history traits and other complex phenotypes that can only be studied from a quantitative perspective. While variation in life-history traits has generally not been considered under stressful conditions, changes in other quantitative traits have been related to periods of environmental stress, such as morphological changes in Darwin's finches in the Galapagos under drought conditions (Boag and Grant 1981).

1.3 Stress responses and measurement

As noted above, stress is characterized by the responses of a biological system to extreme environmental conditions. The effects of an environmental stress on an animal can be outlined by the sequence in Fig. 1.3 which was adapted from Blackstock (1984) and Branch *et al.* (1988) and originally devised for marine invertebrates, although it covers general features applicable to other organisms. This sequence will serve to

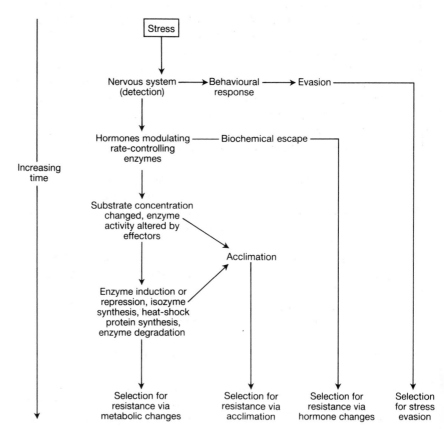

Fig. 1.3. Time related sequence of effects of environmental stress and types of selection relevant to each level of response. (Modified from Blackstock 1984; Branch *et al.* 1988.)

introduce the different aspects of stress responses at the individual and population levels.

Once a stress is detected, organisms may attempt to evade it by a behavioural response. Many animals can respond quickly to a stress by behavioural means. For example, numerous ectotherms change resting sites according to whether they need to increase or decrease their temperature. Organisms may also evade a stress by becoming quiescent or dormant although this is a much slower response that may occur after the detection of a stress or as a reaction to a non-stressful environmental change that signals the advent of stressful conditions. We will refer to both these types of responses as 'stress evasion' because they result in an organism no longer being exposed to a stress.

If behavioural evasion does not take place, then a number of hormone-mediated changes in metabolic processes may occur. The most rapid of these involve changes in enzyme activity, followed by changes in substrates and effectors. Slower metabolic changes involve enzyme synthesis or degradation processes although these can still occur in a few minutes. Heat-shock proteins are synthesized directly after organisms experience one of a wide range of stressors and may be a particularly important component of this response.

Organisms that counter the detrimental effects of a stress with partial or complete recovery are considered to show acclimation in which case the organism has re-established homoeostasis and therefore acquired increased stress resistance. Acclimation, which can occur after exposure to normal ranges of an environmental factor as well as to stressful conditions, refers to changes in resistance to a single factor, as distinct from acclimatization that involves adjustment to two or more factors as is commonly the case in nature. Environmental modification of stress resistance may be of a more permanent form than immediate physiological changes comprising acclimation. One type of change that occurs in some animals is 'developmental conversion' where alternative genetic programs are activated during development and environmentally mediated changes are not reversible (Smith-Gill 1983). Developmental conversion can be distinguished from processes such as acclimation where the environment directly modifies the developmental program and the genetic program is not altered. Both acclimation and developmental conversion represent examples of phenotypic plasticity which describe the variable phenotypic expression of traits by the same genotype under a range of environments.

Finally, there is the possibility of natural selection on genetic variants differing in metabolic efficiency which may occur when there is differential impairment or mortality of genotypes after exposure to a stress. This constitutes selection on genetic variation for stress responses and is the slowest form of response to an environmental stress, occurring at the population level resulting from selection at the individual level. Figure 1.3 emphasizes that genetic variation may exist at several levels, including behavioural evasion, hormone levels and phenotypic plasticity as well as metabolic changes. These genetic changes are the stress responses that are most extensively discussed in this book.

Figure 1.3 indicates that genetic changes in stress resistance may arise from various types of phenotypic changes that help to maintain metabolic homoeostasis. Enzymes or regulatory processes may be altered so that normal rates of metabolic flux are maintained under diverse environmental conditions. There are many examples of such changes from biochemical comparisons of populations or species living in different environments (Hochachka and Somero 1984). Stress resistance may also involve a reduction in the use of resources so that these can be conserved during adverse conditions. Plants occupying continuously unproductive conditions have inherently slow rates of growth and carbon turnover that may lead to resource conservation. Animals can respond to a restricted level of energy intake by a reduction in metabolic rate or (in endotherms) by a reduction in body temperature (e.g. Forsum *et al.* 1981; Chappell and Bartholomew 1981). Numerous phenotypic changes can contribute to resistance by helping to minimize the effects of an adverse environment on the internal environment rather than via metabolic changes. For example, plants close their stomata at times of water stress to maintain the turgor pressure in leaf cells necessary for structural integrity. Intertidal mussels close their mantle cavities when they are exposed to prevent desiccation stress. In insects, the composition of cuticular lipids changes in response to dry conditions to increase the waterproofing properties of the cuticle (Hadley 1977). Genetic differences in stress resistance will result when individuals differ in their ability to carry out these responses. Stress resistance can also be based on inherited morphological traits that influence the internal environment. For example, body colour may affect the resistance of organisms to heat stress by altering internal temperatures and body size can change the susceptibility of individuals to high temperature or low humidity by altering the surface-to-volume ratio.

Organisms that are unable to counter the effects of an environmental change by evasion or resistance will become increasingly stressed, and many methods have been proposed to measure stressed states. These include changes in biochemical variables (glucose, enzyme function, hormone changes, free amino acids), morphology, physiological variables (reproduction, growth, osmotic and ionic regulation, nitrogen excretion), and behaviour (locomotion, orientation, feeding, and avoidance responses). As Ivanovici and Wiebe (1981) point out, many of these methods are limited because of a lack of sensitivity, a need for long exposures to detect responses, inconsistent responses, as well as specificity to species or organisms.

An increase in respiration may be a general early warning sign of a stressed state since repairing damage caused by the disturbance requires increased energy expenditure or the diversion of energy from growth and reproduction to maintenance (Odum 1985). Odum (1983) gives an example of a fish in a pond receiving heated water from industry or a power-plant, which will need to devote metabolic energy to cope with the elevated temperature stress. A consequence will be insufficient metabolic energy for food procurement and reproductive activities required for survival. As extreme conditions are approached, adaptation becomes progressively more costly since progressively more energy is diverted to maintenance (Odum 1985). Extreme conditions including salinity, heavy metals, oxygen deprivation, extremes of heat and cold, and desiccation pose costs for the maintenance of normal protoplasmic homoeostasis and the normal functioning of membranes and enzymes. Respiration has been found to increase in response to numerous stresses, but small increases are often difficult to detect, especially in natural populations, and overall respiration rates may decrease if stress responses involve the conservation of resources.

A biochemical approach to assessing stress experienced by biological systems is to monitor changes in the concentrations of energy carriers (Ivanovici and Wiebe 1981). This approach is generally applicable to a wide range of organisms and is related to changes in the energy balance of cells. One measure is the adenylate energy charge (AEC), being an index calculated from measured amounts of adenosine triphosphate (ATP), adenosine diphosphate (ADP), and adenosine monophosphate (AMP), whereby (Atkinson 1977; Hochachka and Somero 1984):

$$AEC = \frac{ATP + 1/2\,ADP}{ATP + ADP + AMP}$$

The AEC represents the amount of metabolically available energy stored in the adenine nucleotide pool. It also regulates the activities of enzymes involved in processes that generate or utilize ATP as outlined in Fig. 1.4. A cell is expected to be at a steady metabolic state when ATP utilization and regeneration balance (the intersection in Fig. 1.4). AECs can therefore be used to express an organism's deviation from a steady metabolic state (i.e. from metabolic homoeostasis). This provides a measure of stress, which becomes a significant reduction of AEC as a consequence of an environmental perturbation. This reduction, which reflects a drain on metabolic energy resources, renders the organism

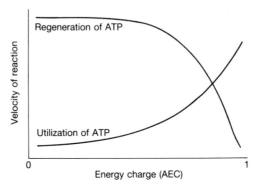

Fig. 1.4. Effect of adenylate energy charge on activities of rate controlling enzymes in reaction sequences regenerating and utilizing ATP. (After Atkinson 1977.)

more vulnerable than hitherto to further environmental change (Ivanovici and Wiebe 1981). Three ranges of AEC are typical in pure cultures and individual organisms:

(1) between 0.8–0.9 for organisms where environmental conditions are optimal or non-stressed, and active growth and reproduction are likely;

(2) between 0.5–0.7 for organisms that are under recognized limiting or non-optimal conditions, and where growth rates are slower, reproduction is greatly reduced or does not occur, but viability is retained so that on a return to normal conditions, they resume all of the characteristics of the previous state;

(3) about 0.5 for organisms that are under severe stress and viability losses occur even after a return to control conditions.

According to Ivanovici and Wiebe (1981), the advantages of the AEC as a stress measure are:

(1) it correlates with physiological condition or growth rate in a diversity of organisms;

(2) the universal role of adenine nucleotides in energy metabolism of all living organisms means that measures are obtainable across the biological spectrum;

(3) variation in AEC is less than for individual nucleotides or physiological measures;

(4) response times are fast.

Furthermore, the AEC decreases for a wide range of environmental perturbations such as reduced salinity, anoxia, increased temperature, hydrocarbon contamination, glucose depletion, desiccation, and nutritional limitation. For example, Din and Brooks (1986) found that the AEC ratio is a reliable index for determining sub-lethal and chronic effects of chemical industrial wastes on phytoplankton. Nevertheless, there are some examples where the AEC is conserved during stress and AECs stabilize at a minimum value (e.g. Wijsman 1976). Absolute measures of AECs may therefore not always be indicative, and the initial rate of decline of AECs may be a more useful general indicator. Some examples of changes in AECs in response to stress are given in Blackstock (1984) and Van Waarde (1988).

We have used the term 'resistance' rather than 'tolerance' in this section, even though these terms are used interchangebly in much of the literature on environmental stress. Levitt (1980) argued that resistance is the more general term and that tolerance should be restricted to those cases where organisms can prevent, decrease or repair the strain resulting from a stressor. Levitt suggested the term 'stress avoidance' where the effects of a stress are prevented from reaching tissues.

1.4 This book

The emphasis of this book is on the extreme end of the stress gradient at the limits of resistance. The persistence of an environmental stress at this level will result in non- reversible damage and mortality irrespective of the density of organisms. Such situations are said to represent hard selection in the sense of Wallace (1981), as distinct from soft selection where the fate of an individual depends upon the phenotype and genotype of its immediate neighbours and there is density and/or frequency dependence arising from competitive interactions.

We believe that a major emphasis on environmental stress is justified for a number of reasons. First, there are many evolutionary and ecological situations where responses to environmental stress appear to be important (Chapter 2). Many of the field examples of natural selection reviewed by Endler (1986) involve genetic changes in populations experiencing extreme stress levels. These include selection on body size in Darwin's finches, shell colour patterns in gastropods, and melanism in insects. Major periods of extinction in the fossil record appear to be associated with periods of climatic stress. The distribution of many species is determined by climatic extremes and the absence of resistance to stress appears to limit range expansion by species. Many recent cases of extinction and changes in species distributions are the result of stresses related to human activities, and these will become more important and widespread assuming the predicted advent of global warming.

Secondly, many specific phenotypes appear unique to stressful situations and cannot be studied under non-stressful conditions. At the molecular level, there are heat-shock proteins and other molecules that are only expressed when organisms are stressed. At the organismic level, there are numerous physiological and morphological traits which enable organisms to survive under stressful conditions but do not increase survival under optimal conditions.

Thirdly, characteristics that increase stress resistance or evasion may have a cost due to factors such as the metabolic energy required for their development and maintenance. This suggests that phenotypes favoured under stressful conditions may not be favoured under non-stressful conditions. Genes influencing stress resistance may therefore have pleiotropic effects which could underlie interactions between traits as well as forming the basis of trade-offs between environments (Chapter 7).

Fourthly, stress responses are involved in interactions between life-history traits. Trade-offs will often occur among life-history traits because physiological constraints mean that increased performance for one trait will often be accompanied by decreased performance for another trait. The degree of stress at different life cycle stages will determine the nature of trade-offs between traits, and optimality models can be used to predict life histories resulting from these trade-offs (e.g. Sibly and Calow 1989).

Fifthly, the nature of natural selection under stress conditions differs from that under more optimal conditions because of the high intensity of selection and the short, intermittent nature of stress periods. This means that modelling selection for stress traits requires short periods of high selection intensities rather than continuously weak selection assumed in many models of selection on quantitative traits (e.g. Lande 1975, 1976). This can have implications for the genetic basis of traits selected under stressful conditions because major genes may be favoured when selection intensities are high.

Finally, stress response traits have a number of inherent advantages over other traits that are studied in evolutionary genetics. The traits are easily defined and measured in simple laboratory assays. This means that they are suitable for genetic analysis. They can be directly related to natural selection. Responses to physical stresses that result in hard selection do not require an understanding of complex biotic interactions that are associated with many other forms of selection.

Perhaps the major difficulty in studying the evolutionary genetics of stress response traits is that periods of extreme environmental stress may not be particularly common in many populations, implying an emphasis upon studies of relatively rare events in the wild. Extreme stresses may be more common in populations at species margins and human activities provide several cases where specific stressors have been applied frequently, such as the application of pesticides and herbicides for the control of insect pests and weeds. However, the emphasis of much research in evolutionary biology has been on relatively benign situations, including most research on evolutionary genetics in laboratory environments. Comparative studies of populations have generally not involved populations from marginal habitats. We are therefore often restricted to considering genetic data from populations that have not been exposed to environmental stress, or to interspecific comparisons. These problems are also encountered by ecological physiologists who usually compare populations or species living in different environments to study the mechanisms associated with adaptation to stresses, and only rarely consider species under unfavourable conditions (Osmond *et al.* 1987).

Our emphasis on genetic changes under extreme environmental stresses and their evolutionary consequences contrasts with that of most discussions on environmental stress which are more concerned with documenting the nature of physiological and biochemical adaptations of organisms in extreme environments. For example, ecophysiologists have been concerned with describing the way that organisms are adapted to the environment in which they live, and an enormous literature has built up on physiological mechanisms associated with stress resistance in both plants and animals. Most studies have involved comparisons of species living in different environments. In considering physiological aspects of stress resistance, we (unlike physiological ecologists) are interested less in describing adaptations at the species level than in examining physiological differences between populations and between individuals from the same population. This does not mean that findings from interspecific comparisons are irrelevant because traits possessed by organisms occupying extreme environments may provide indications of the types of traits possessed by the more stress-resistant individuals within a population. Some of our discussion comes from the interspecific level simply because there may be little information on genetic variants within populations.

The main objective of this book is to examine those genetic aspects of responses to environmental stress that may be generally applicable to a range of organisms. At the biochemical and physiological level, we look for common mechanisms that are selected when populations are exposed to different environmental stresses. The types of genetic variation that have been related to stress responses are examined, focusing particularly on quantitative genetic variation. The effects of stress on heritability and the genetic architecture of stress response traits are considered. Finally, we examine the nature of interactions between stress response traits and other traits influencing fitness in non-stressed situations. These discussions will form the basis for a consideration of genetic conservation and explanations for limits to range expansion by species. Throughout the book we will make attempts to link the genetic level with the phenotypic level and the statistical approach of quantitative genetics to physiological and biochemical considerations. We believe that increased understanding of evolution under stress will come from combining investigations at these different levels. We will emphasize the importance of an integrated approach to genetic variation in stress responses that may be partly based on the metabolic costs imposed by extreme conditions, and general metabolic changes that enable organisms to overcome the effects of an environmental stress.

1.5 Summary

Our definition of stress emphasizes changes in a biological system as a consequence of exposure to extreme physical conditions in the environment. Physical factors can cause stress directly, while biological factors such as competition may affect stress levels indirectly by interacting with physical factors.

Darwin regarded organism–environment interactions as central for an understanding of evolution. He emphasized competition, but recognized the importance of climatic stresses at the extremes of species distributions. The evolutionary consequences of extreme stress were occasionally discussed by other evolutionary biologists, in particular Schmalhausen. Older quantitative studies on resistance to extreme stresses revealed substantial genetic variability within and between populations. Recent approaches to evolutionary genetics have occasionally emphasized stress in cases such as heat-shock proteins or differences between enzyme variants under extreme conditions. A recent resurgence in quantitative genetics has focused on life-history traits in conditions that are generally non-stressful.

Organisms may respond to stress by phenotypic plasticity involving rapid reversible changes or more permanent changes in development. Genetic changes occur more slowly at the population level. At the energetic level, stress has a metabolic cost which can be detected by biochemical indices such as the adenylate energy charge (AEC).

Stress may have a major impact on many evolutionary and ecological processes, and specific traits and genetic changes may be associated with evolutionary changes under stressful conditions. The importance of stress in natural populations may have been underestimated because stressful periods are often assumed to be rare or to predominate at species borders that are infrequently studied in evolutionary biology.

2. The evolutionary and ecological importance of environmental stress

Periods of extreme environmental stress have been emphasized in many areas of the evolutionary and ecological literature. We briefly describe some of these areas to give an overview of the biological importance of environmental stress, including patterns in the fossil record, the impact of unusual climatic conditions on populations and the association between climatic extremes and species distributions. Attempts to classify organisms by the nature of their habitats also incorporate stress, and these are briefly discussed along with the interaction between stress and competition.

2.1 Fossil extinction and environmental stress

The fact that virtually all plant and animal species that have ever lived on earth are now extinct implies a major role for extinction in the evolution of life (Raup 1986). In the last 600 million years, there have been five major mass extinctions, around 440, 365, 250, 215 and 65 million years ago (Fig. 2.1). In the late Permian event 250 million years ago an estimated 96 per cent of the existing species were killed. The terminal Cretaceous mass extinction 65 million years ago is of particular interest because of its likely origin in a comet or asteroid impact (Alvarez *et al.* 1980). Even so, whatever the initial precipitating event, climatic deterioration and sea level changes are the most commonly cited immediate and ultimate causes of mass extinctions.

A recent analysis of 19897 fossil genera (Raup and Boyajian 1988) summarized in Fig. 2.1 indicates that for nine groups of marine organisms, most classes and orders show largely congruent rises and falls in extinction intensity throughout the Phanerozoic. Regardless of the mechanism, the taxonomic as well as the ecological groupings show approximately the same extinction profiles. This means that the extinction events cut across functional, physiological, and ecological lines, suggesting common causes, and indicates that they are physically rather than biologically driven. Raup and Boyajian (1988) also considered only reef organisms that share a similar habitat but cut across taxonomic lines, and found that the extinction pattern in this group was similar to

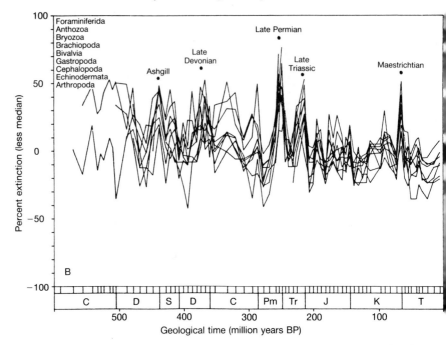

Fig. 2.1. Extinction profiles of nine major Phanerozoic groups. (After Raup and Boyajian 1988.)

that found in the marine biota as a whole. These results indicate a parallel history of environmental stresses in the determination of extinctions. A further departure from earlier gradualistic explanations of extinction comes from a number of analyses (Raup and Sepkoski 1984) indicating that extinction events are generally spaced in geological time with periods ranging from 26 to 32 million years. If periodicity can be established, a common cause is virtually essential.

It is an interesting commentary on scientific fashions that Raup and Boyajian (1988) found the message of physical causes being put forward over 30 years ago by other workers who had noted similarities in the extinction records of quite disparate groups. For example, Newell (1952) concluded that times of low evolutionary change tend to coincide in a large proportion of the major groups. This conclusion was necessarily based upon far fewer records than those used by Raup and Boyajian (1988).

Extinction has been viewed by some as a constructive force leading to an improvement in the mean adaptive level of the whole biota via the elimination of some poorly adapted organisms. However, the mass extinctions and periodicities indicate that extinction is a selective but not constructive process. As Raup (1986) writes:

If mass extinctions are the result of environmental stresses so rare as to be beyond the 'experience' of the organisms, extinction may be just a matter of the chance susceptibility of the organisms to these rare stresses.

He cites the hypothetical (but not impossible) example of large doses of ionizing radiation which, in terrestrial habitats, could cause extinctions among mammals but would have negligible effects on most insects and plants. There would then be highly selective extinction, but one without constructive effects because periods of stress would need to be common for the evolution of stress- resistant genotypes by natural selection.

The exact nature of environmental stresses leading to extinctions are unknown. Many lines of evidence indicate that temperature is a prominent proximal cause of marine mass extinctions (Stanley 1984) irrespective of the triggering event. A persistent theme of mass extinctions has been the decimation of tropical marine biotas, with less severe losses occurring at high latitudes. For many mass extinctions there is physical or biological evidence of contemporaneous climatic cooling independent of the latitudinal pattern of extinction (Stanley 1984).

2.2 Acyclic climatic stress

To what extent are individuals in extant natural populations exposed to environmental stress? Many populations experience diurnal or seasonal periods of unfavourable climatic conditions, adaptation to which is generally expected to involve biological clocks or other time-measuring devices. The expression of such rhythmic responses is mediated via the neural and hormonal systems and may allow organisms to evade stress. In other words, there is an inherent ability of organisms to measure time, so providing a level of predictability which they can use to evade a stress by mechanisms such as migration or entering a dormant phase.

In contrast, unfavourable conditions may also be acyclic, and these do not provide advance warning which would permit physiological and/or behavioural alterations. It is these events that can result in major evolutionary changes, including population and species extinctions, and they are often described as ecological crises. Extinction may be avoided if acyclic perturbations of equivalent magnitude to the cyclic changes can be overcome by the physiological and biochemical adaptations that have evolved to counter the cyclic changes, or if some genotypes in the base population can withstand the stressful conditions so that populations evolve increased resistance.

A feature of acyclic periods of climatic stress is that their severity and occurrence is extremely variable. This can be illustrated by the soil moisture deficits during growing seasons for a 279-year period at Kew, England (Fig. 2.2). Moisture provides a fairly direct assessment of the degree of drought stress on the vegetation and associated fauna, and should be closely tied to productivity and resource levels. There is no apparent regularity in the occurrence of severe stress years. Predicting ecological crises is therefore exceedingly difficult. Crises are events of short duration relative to the total time under consideration, but they are times when large biological changes may occur. The problem is that climatic changes may approach the level of an ecological crisis only once every 50–100 generations, and studies of much longer time span than is usual in ecological studies are necessary for their documentation (Raup 1981; Wiens 1977).

There is a great deal of evidence that attributes extinctions and large reductions in population numbers to these acyclical stress periods. As discussed in Chapter 1, Darwin (1859) recognized the occurrence of stress, but he tended to view evolution as a more gradualistic process emphasizing the intricacies of interactions among species. He nevertheless described some cases of acyclic climatic stress. For example:

the winter of 1854–5 destroyed four-fifths of the birds in my own grounds; and this is a tremendous destruction, when we remember that ten per cent is an extraordinary severe mortality from epidemics with man.

Numerous other examples of acyclical stress periods can be found in the literature. In marine invertebrates, periods of thermal stress due to low and extremely high temperatures have often been implicated in high mortality rates, as discussed in Kinne (1971). A number of marine and brackish-water invertebrates cannot withstand freezing, and massive mortality may be associated with extreme conditions. For example, the

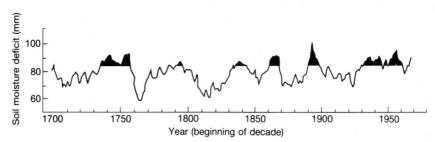

Fig. 2.2. The soil moisture deficit at Kew, England, averaged over the growing seasons of a 279-year period using 10-year running means. An arbitrary level of 84 mm is shown to accentuate periods of higher deficits (Wigley and Atkinson 1977). There is no apparent regularity in the occurrence of severe years.

extreme cold winter of 1962/63 led to death rates of up to 90 per cent in *Branchiostoma lanceolatum* near Helgoland (North Sea), and high mortalities were also recorded among a number of marine invertebrate species in the North Sea and along the British coasts. The immediate circumstances leading to cold death in the sea are complex (Kinne 1971). Direct lethal damage of body structures seemed to be largely restricted to cases involving ice formation. More importantly, there were critical reductions in life-supporting functions such as motility, respiratory movements, and muscular strength, as well as in the coordination and integration of body functions, and there was a decrease in resistance to parasitic and microbial infections. Such collective stresses did not lead to immediate death, which tended to occur at maximum rates weeks or even months later. At the other end of the scale, many marine invertebrates are killed by temperatures not far above those experienced in their natural habitats, and several instances of mortality due to high temperatures are given in Kinne (1971).

Mass mortalities of fish in the sea have been linked to various disturbances including volcanic eruptions, seaquakes, waterbloom, severe storms, and sudden changes in salinity and temperature (Brett 1970). Fish death from extreme temperature has been well documented, including episodes of massive death associated with the cold winter of 1938–40 along the southern tip of Florida, USA, and the 1962–63 winter in the North Sea.

A recent example of an ecological crisis in marine environments is the El Niño event of 1982–1983, which resulted in sea warming and decreased nutrients in tropical and subtropical latitudes of the Pacific Ocean; its biological impact has been reviewed by Glynn (1988). Ecological damage included coral death ranging from 50 to 98 per cent due to high temperature, and extinction of entire reefs occurred in the Galapagos. Benthic algal populations suffered from nutrient stress, high temperature stress and physical damage from storms. Numerous sea birds declined in abundance due to food shortages, and total reproductive failure occurred in many populations. Decreases in abundance also occurred in marine iguanas, fur seals, and, in particular, in sea lions where some age classes showed almost 100 per cent mortality. In addition, biotic factors produced further reductions in the abundance of some populations. For example, blooms of dinoflagellates caused coral mortality, and grazing by sea urchins caused damage to coral reefs and prevented the re-establishment of algae.

There are several records of ecological crises in terrestrial invertebrates. During an investigation of the dynamics of two species of the checker spot butterfly, *Euphydryas*, at the time of a major Californian drought in 1975–77, several populations became extinct, some were dramatically reduced, others remained stable, and at least one increased (Ehrlich *et al.* 1980). The unusual spring weather climaxing in a late June snowstorm in subalpine Colorado led to extensive damage to herbaceous perennial plants and reduced the size of insect and small mammal populations (Ehrlich *et al.* 1972) whereby at least one butterfly (*Glaucopsyche lygdamus*) population became extinct because of the destruction of the inflorescence of a common perennial lupin, *Lupinus amplus*. Another lepidopteran example is the high mortality of Monarch butterflies following a severe winter storm in January 1981 in Mexico (Calvert *et al.* 1983). There are numerous reports of changes in the abundance of insect pests in stressful conditions. In particular, pest outbreaks associated with climatic extremes have been widely documented because of their potential agricultural importance (e.g. Wallner 1987).

Acyclic climatic changes have been associated with numerical changes in populations of terrestrial vertebrates. For example, severe weather associated with low temperature and unusual spring precipitation led to widespread mortality and local population extinctions in cotton rats (*Sigmodon hispidus*) in Kansas (Sauer 1985). A severe drought resulted in 85 per cent mortality among Darwin's finches (*Geospiza fortis*) in the Galapagos (Boag and Grant 1981). In Australia, examples include the large-scale effects of a drought between April 1982 and March 1983 that led to a 40 per cent decline in kangaroo numbers over an area of more than one million square miles (Caughley *et al.* 1985), and the more localized effects of a heat wave and drought conditions in south-western Queensland which led to 63 per cent mortality among a koala population (Gordon *et al.* 1988).

Plants often occur under conditions unfavourable for growth and reproduction, and may commonly be under stress in natural environments. Most studies in plant physiological ecology have compared the physiological responses of related plants to stress from locations characterized by variation in a single environmental factor. However, Osmond *et al.* (1987) emphasize that covariation and interaction among major stress factors are important in determining the performance and distribution of plants along gradients. Figure 2.3 illustrates the way in which stress factors are likely to interact in different ecosystems from the Northern Hemisphere. The most prominent stresses are temperature and water availability. Superimposed upon these two major axes are complex gradients arising from other abiotic factors such as light intensity and pollutants, and biotic factors such as herbivory, disease, and competition. The importance of biotic factors will often depend indirectly on stressful conditions: for example, outbreaks of phytophagous insects may be associated with drought or other environmental

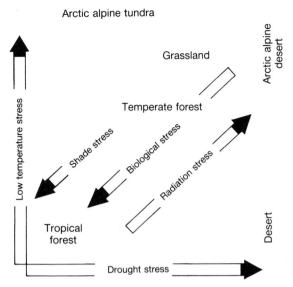

Fig. 2.3. Schematic gradients of environmental stress in relation to major vegetation types. The x-axis is a gradient of decreasing temperature, and the y-axis represents decreasing water availability. (After Osmond *et al.* 1987.)

stresses (Mattson and Haack 1987). The incidence of bark beetles and leaf feeders typically increases after periods of unusually warm, dry weather. Stress conditions for plants appear to lead to optimum conditions for the insect, although severe and prolonged drought may become debilitating for the insect.

The effects of physical factors on plant production are even apparent in agricultural systems, where attempts are made to optimize conditions. Boyer (1982) gives several relevant figures. Comparing the yields of major US crops, record yields were three to seven times the average yields, although for crops having marketable vegetative structures (potatoes and sugar beet), the discrepancy was lower (three times) than for the other crops because the complexities of reproductive development are not involved. In analysing the causes of the discrepancies, disease and insect losses were minor (4.1 and 2 per cent, respectively) by comparison with weed competitors, inappropriate soils, and unfavourable environments. The kinds of environmental factors that affect plants adversely can be quantified, even though they overlap and are often sporadic in their impact. Table 2.1 shows that drought, i.e. low water availability, is by far the most important problem, while excessive wetness and cold are also of major significance. Only 12.1 per cent of the land surface under consideration is free from problems associated with

Table 2.1. Area of the United States with soils subject to environmental limitations of various types. (Data from Boyer 1982.)

Environmental limitation	Area of US affected (%)
Drought	25.3
Shallowness	19.6
Cold	16.5
Wet	15.7
Alkaline salts	2.9
Saline or no soil	4.5
Other	3.4
None	12.1

physical factors and this area is the most productive when supplied with plant nutrients to replace those removed when crops are harvested.

Agricultural studies have focused on variation in factors that normally limit productivity, but the importance of extreme climatic conditions have been recognized in some cases. Parry (1978) discussed the effects of severe weather during the Little Ice Age on agricultural limits in southeastern Scotland, where at the margins of agricultural viability through cold or dryness,

(1) extreme events occur relatively frequently;

(2) the frequency of such events may change dramatically as a result of even a minor change in mean climate;

(3) the chance of two successive extremes is even more sensitive to changes in the mean.

Parry showed that the area of land cultivated and the settlement pattern in this region was largely determined by climatic changes, land and settlements being abandoned as the climate became colder. Parry described similar findings from other areas of marginal farmland in Europe. While conceived in terms of an agricultural model of the cold marginal uplands of northern Europe (Parry and Carter 1985), extremes of climate may be generally important in an agricultural context. For example, Mearns *et al.* (1984) presented evidence for an effect of temperature extremes on corn yield in the U.S. corn belt. Runs of five consecutive days of extreme heat substantially reduced yield because this period was long enough to affect the entire temperature-sensitive silking stage. It is important to consider the potential effects of the temperature increase proposed for the greenhouse effect in this light, especially as quite minor temperature shifts within the increased temperature ranges that have been proposed can have major effects upon maturing grain crops, and hence upon agricultural yields (Schneider and Lander 1984).

We have only focused on climate in this section, but there are numerous cases of large changes in abundance and population extinctions in response to temporal variation in other environmental stresses, in particular the many pollutants associated with human activity. Heavy metals in soils and water and atmospheric pollutants can have a marked influence on the abundance of species (e.g. Bradshaw 1984; Hutchinson 1984), and their effects will become increasingly widespread as the impact of human activities becomes more apparent. It is appropriate to summarize this section by quoting Darwin (1859):

In the case of every species, many different checks, acting at different periods of life, and during different seasons or years, probably come into play; some one check or some few being generally the most potent; but all will concur in determining the average number or even the existence of the species.

2.3 Species distributions

Climatic variables are important determinants of limits to species distributions in the absence of abrupt geographic barriers. Australia forms an excellent case study, with its wide array of climatic types. To this end an Australian bioclimatic prediction system (BIOCLIM) has been developed from twelve parameters derived from the climate (Busby 1986a) which characterize annual, seasonal and extreme climatic components (Table 2.2). From known distributions of species, the BIOCLIM system is used to determine the bioclimatic envelopes of

Table 2.2. Climatic characteristics used in the BIO-CLIM analysis of Busby (1986a)

(1) Annual mean temperature
(2) Minimum temperature of coldest month
(3) Maximum temperature of hottest month
(4) Annual temperature range $(3-2)$
(5) Mean temperature of the wettest quarter (3 months)
(6) Mean temperature of the driest quarter
(7) Mean annual precipitation
(8) Precipitation of the wettest month
(9) Precipitation of the driest month
(10) Annual precipitation range $(8-9)$
(11) Precipitation of the wettest quarter
(12) Precipitation of the driest quarter

species based upon climatic similarities with actual distributions. One of the major aims has been to determine the distributional limits of species, and it has been successfully used in determining the distribution of elapid snakes, the rain forest tree *Nothofagus cunninghamii*, kangaroos, and chromosomal taxa of the Australian grasshopper, *Caledia captiva* (F) (Busby 1986*b*; Nix 1986; Caughley *et al.* 1987; Kohlmann *et al.* 1988).

As an example of this approach, Caughley *et al.* (1987) considered the distribution of the eastern grey kangaroo *Macropus giganteus*, the western grey (= southern grey) kangaroo *M. fuliginosus*, and the red kangaroo *M. rufus*. These are three of the largest species of the Macropodoidea (kangaroos, wallaroos, wallabies and rat kangaroo). The range of each overlaps to some extent with the other two, but as a generalization *M. giganteus* lives in the east of Australia, *M. fuliginosus* occurs across the south of the continent, excluding Tasmania, and *M. rufus* lives in the arid and semi-arid zones that occupy much of the interior, but which also reach the coast in Western Australia. Using the BIOCLIM system, Caughley *et al.* (1987) found that the degree of sympatry and allopatry appears to be determined by the independent reaction of each species to specific and differing climatic stimuli, rather than by biological interactions between species. Figure 2.4 provides a discriminant analysis of habitat characteristics occupied by the three species. The species have a common range in the climatic zone characterized by low seasonality of rainfall and a semi-arid mean annual precipitation and temperature. In the discriminant analysis, the x-axis features annual mean temperature, maximum temperature of the hottest month, minimum temperature of the coldest month, and mean annual precipitation, and the y-axis features annual mean temperature, precipitation of the driest quarter and temperature of the driest quarter. Climatic variables representing extremes are therefore closely correlated with the distributions of these species.

The association between climatic extremes and species distributions has been emphasized in numerous studies outside the Australian continent. Woodward (1987) has outlined evidence that climate is the predominant factor influencing the world-wide distribution of plants. In particular, he argued that low temperature extremes controlled the distribution of several vegetation types, while local climatic conditions and soil type were also important. In birds, large, unpredictable variations in climate were considered to be the most important factor affecting the number of species in North American grassland communities (Wiens 1974). Root (1988*a*) found that the range of bird species in winter was closely associated with the mean minimum temperature: more than half the birds in North America have their northern

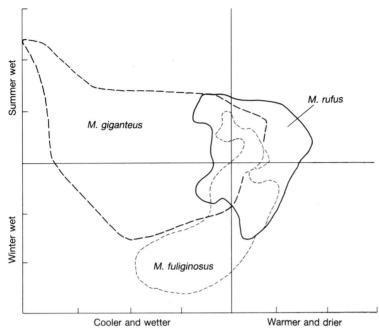

Fig. 2.4. Distributions of three species of kangaroos within the space defined by two discriminant functions. The first (x-axis) features annual mean temperature, maximum temperature of the hottest month, minimum temperature of the coldest month, and mean annual precipitation. The second (y-axis) features annual mean temperature, precipitation of the driest quarter, and temperature of the driest quarter. (After Caughley *et al.* 1987.)

range border during winter associated with a particular isotherm, which may also reflect a physiological limit for thermoregulation. McNab (1973) found that the distribution of vampire bats in Mexico followed the 10°C minimal isotherm for January and argued that this reflects the bat's limit for thermoregulation because of the maximum meal size that it can transport.

A limitation of many of these studies is that correlations between species distributions and environmental variables may not represent a causal relationship, although several lines of evidence suggest that distribution limits are determined by environmental extremes because findings have been used in a predictive sense. In the process of 'climatic matching', information about the origin of species or strains can be used to predict their likely success in a new environment. For example, Burt *et al.* (1976) obtained climatic information for the places of origin of 158 accessions of the tropical pasture legume, *Stylosanthes*, into Australia, and showed clearly that predictions of agronomic performance could be made from climatic data. Burt *et al.* used a profile of climatic characteristics emphasizing extremes in their analysis (Table 2.3), and concluded that if a legume is needed in an area for which no known accession is suitable, it should be possible to specify the climate of origin, and hence the geographical region, in which such a legume should be sought. These climatic characteristics have similarities with those used in the BIOCLIM analysis (Table 2.2).

The importance of responses to environmental stress in determining species distributions can also be examined experimentally by relating species distributions to the results from 'stress tests' in the laboratory or in the field in which organisms are subjected to a range of experimental conditions. Such a physiological approach has helped ecologists to understand the distribution of organisms in nature (Odum 1983), although interactions between individuals may influence stress resistance and intact ecosystems need to be considered together with experimental laboratory studies, which necessarily isolate individual organisms from populations and communities. The actual range of resistance in nature tends to be narrower than the potential range as elucidated under laboratory conditions.

As an example of this approach, Forman (1964) grew the moss *Tetraphis pellucida* under various combinations of controlled temperature, relative humidity, pH, and light intensity in a microphytotron. This species occurs on decaying wood in cool, moist coniferous forests of the north temperate zone. Forman found that temperature and relative humidity alone were inadequate to explain the continental distribution of the species from experimental growth data. Mean monthly maximum and minimum temperatures were more useful than mean monthly or annual temperatures, emphasizing the importance of extremes, and these variables had to be combined with corresponding relative humidities to obtain reasonable correlations between the actual distribution and results from the growth experiments.

Table 2.3. Climatic characteristics used in the analysis by Burt *et al.* (1976)

1. Latitude
2. Altitude
3. Number of months giving 85–92% of total rainfall
4. Highest monthly maximum temperature during wet season
5. Lowest monthly minimum during wet season
6. Lowest monthly maximum in dry season
7. Lowest monthly minimum in dry season
8. Average annual rainfall

Another example of this approach is provided by the genus *Drosophila* which is believed to have evolved in the tropics and spread to the temperate zone. This range expansion implies a climatic shift from a uniformly warm humid environment to a seasonal pattern with greater extremes of temperature and humidity. This shift is associated with correspondingly higher resistance to temperature and desiccation stresses in the laboratory (Fig. 2.5). Among the endemic fauna of Australia, three species, *D. enigma*, *D. lativittata* and *D. nitidithorax*, are attracted to fermented-fruit baits and have spread into orchards from natural habitats. The colonization potential of *D. enigma* is further confirmed by its recent discovery on the North Island of New Zealand, where the North American species, *D. pseudoobscura*, has also been found. All of these species are more resistant to the two interacting stress factors in Fig. 2.5 than tropical species, and so clearly have ecological phenotypes consistent with range expansion (Parsons 1982*a*, 1987*a*). *D. inornata* is a temperate-zone Australian endemic species occasionally found in urban regions. It is mainly found in permanently moist

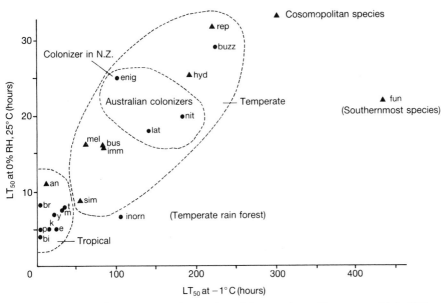

Fig. 2.5. LT_{50} values expressed as a plot of the number of hours at which 50% of flies (sexes combined) died of $-1°C$ stress vs 0% RH 25°C stress for various Drosophila species: an, *ananassae*; bi *bipectinata*; br, *birchii*; bus, *busckii*; buzz, *buzzatii*; e, *erecta*; enig, *enigma*; fun, *funebris*; hyd, *hydei*; imm, *immigrans*; inorn, *inornata*; k, *kikkawaii*; lat, *lativittata*; m, *mauritiana*; mel, *melanogaster*; nit, *nitidithorax*; p, *paulistorum*; ps, *pseudoobscura*; rep, *repleta*; sim, *simulans*; t, *teissieri*; y, *yakuba*. (Adapted from Parsons 1982*a*.)

temperate rain forests, so its sensitivity to desiccation is consistent with its habitat, although it is more resistant than other species completely restricted to rain forests, consistent with its occasional presence in urban collections in winter and spring. The resistance of *D. inornata* to cold is expected since it experiences winters when temperatures are below 10°C for long periods. In summary, the position of a species in the plot in Fig. 2.5 can be related to its habitat, indicating that measures of extreme stress are important in determining species distributions. It is instructive to note the parallels between Fig. 2.3 where environmental stresses of low temperature and drought are shown to be the major stress gradients influencing plant growth and reproduction, and those for *Drosophila* in Fig. 2.5, a genus which depends upon plants for resources.

A third approach to the use of information about species distributions in a predictive manner is to associate historical changes in species distributions with climatic changes. Well-documented examples include the effects of climatic change on the borders of tree populations. Brubaker (1986) reviewed three cases where population borders have responded directly to climatic change. The northern limit of *Picea mariana* in Quebec became more restricted after the onset of cold conditions in the Little Ice Age because a fire killed most trees and the colder temperatures prevented re-establishment. In Sweden, the treeline of *Betula pubescens* rose between 1915 and 1975 in response to higher temperatures, while recent tree establishment in subalpine meadows of ranges in western North America was also associated with warmer temperatures. Additional examples of changes in climate that correspond to changes in the distribution of plant species are given in Woodward (1987).

Collectively, both the experimental approaches and those based upon the development of climatic envelopes and prediction systems emphasize that species distributions are closely associated with physical features of the environment which incorporate climatic extremes. As in the case of abundance, species distributions will also be affected by other abiotic factors such as pollutants. For example, the distribution of lichens can often be related to air pollution, with most species disappearing after the onset of pollution, and a few resistant species increasing their range dramatically (Hawksworth and Rose 1976).

2.4 Stress, habitat classification, and competition

Several authors have classified habitats into quadrangular and triangular representations (Southwood 1977, 1988) in the hope of finding a habitat template of general utility for classifying ecological communities and the types of interactions between species. It is assumed that habitat determines the selection processes acting on species so that the traits exhibited by species and the nature of community interactions will reflect their habitat. Environmental stress forms a central component of these classifications. A template incorporating and summarizing the main features of the various representations from Southwood (1988) is given in Fig. 2.6. The template axes are classified into:

(1) Disturbance axis, representing the frequency of disturbances. Temporary habitats may arise from ecological perturbations ranging from violent and catastrophic events, to lesser events such as the fall of a tree in a rainforest which leads to a series of successional changes. The creation and elimination of temporary pools following a drought provide additional examples.

(2) Adversity axis, representing the degree of stress experienced by organisms in a habitat, due to extremes of various physical factors. The environmental stresses have to be countered by an organism and presumably have metabolic costs.

In this classification, biotic agents have their greatest impact in permanent and undisturbed habitats. These are stable habitats such as tropical rain forests, where productivity and interactions with other organisms, including competition, are expected to be high. More instances of complex interactions between species are expected to occur in the tropics than in temperate systems. Under stable conditions, population size variation should be sufficiently dampened down so that size depends upon competitors, parasites, and predators, which all involve frequency dependent processes (Futuyma 1979; Clarke 1979).

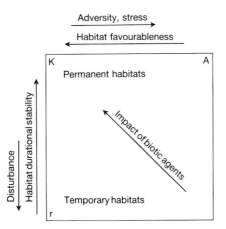

Fig. 2.6. Southwood–Greenslade habitat template. (After Southwood 1988.)

The symbols r and K in Fig. 2.6 come from the r–K selection continuum of MacArthur and Wilson (1967). This continuum represents selection in relation to density-dependent resource availability, where K represents environmental stability and a population near its equilibrium density in which competition is potentially important, while r represents a population in a disturbed habitat and not at equilibrium in which high reproductive rate will be favoured. The third symbol, A, comes from adversity selection (Greenslade 1983) when the environment is predominantly unfavourable and restrictive, only occasionally becoming sufficiently favourable to permit populations to increase. In terms of r, K, and A, stress is primarily associated with A, but also r selection.

Stressful conditions may decrease the importance of competition by reducing the stability of a population. Competition is considered important in habitats where the effects of abiotic stresses are minimal; it should therefore be at a maximum under circumstances of environmental stability, rather than under the more fluctuating environments that tend to cause populations to experience frequent periods of stress, especially at the margins of species distributions. Where stressful conditions occur, populations may tend towards r selection and start to fluctuate in size, so that competition no longer has an effect on size. This may involve the direct effects of climatic stress or indirect effects via the outbreak of disease, predators or herbivores (Coakley 1988; Mattson and Haack 1987). Where environmental stress is persistent, abundance may be low, so that there are few interactions between related organisms. Environmental stress may therefore decrease competition by generating instability and also by decreasing abundance so that potential competitors no longer interact.

These considerations raise the possibility of trade-offs between the performance of species in the different habitat types, if it is assumed that adaptation to one habitat is associated with decreased adaptation to another. This could result in a negative correlation between resistance to adverse conditions and competitive ability at the species level. In other words, competitive ability is traded off against adversity resistance because species selected in stable habitats will have a high competitive ability while species selected in stressful habitats will have a low competitive ability. Such a negative association has been documented by Wilson and Keddy (1986a,b) in a study of seven plant species along a gradient of exposures on the shore of a Canadian lake. Species from sheltered, nutrient-rich shores had high competitive abilities, whereas species from exposed, nutrient-poor shores had low competitive abilities when they were grown together in pairwise combinations. This trade-off has also been discussed at length by Grime (1979) and will be considered further in Chapter 6.

Not all researchers consider that stress resistance and competitive ability are selected in different types of habitats. Stress can increase competitive interactions when populations initially have abundant resources and there is little competition (e.g. Wiens 1977; Connell 1983). An external stress such as a climatic change may decrease the availability of resources to such an extent that individuals have to compete. An example is the reduction in food supply as a consequence of climatic deterioration. In such cases, competition may increase when populations start from a position where they have not saturated the environment (i.e. there is little competition), such as an *r*-selected population in an expansion phase.

As discussed by Welden and Slauson (1986), these points of view may be partly reconciled by separating the importance of competition from the intensity of competition. Demonstrating intense competition in the field does not imply that competition is important, because this can only be determined by comparing the effects of competition with the effects of abiotic factors. Thus while competition may be more intense under stressful conditions, it does not follow that competition will be relatively more important than abiotic factors under such conditions. Persistence of stressful conditions will result in selection for increased stress resistance rather than competitive ability once stress has a direct influence on abundance, and the importance of competition will decrease as direct effects lead to a decrease in abundance. In terms of food supply, selection for starvation resistance will only be intense when food becomes limiting, so competitive ability and starvation resistance may be selected under the same conditions, but starvation resistance will become progressively more important as the food supply continues to be diminished. The intensity of competition will not always increase when adverse conditions are encountered, such as when stress levels increase rapidly. For example, Caughley *et al.* (1985) found that the mortality of kangaroos from starvation during a drought was independent of initial densities, suggesting that the impact of the drought was too rapid to be mediated by competition from a reduced food supply. In addition, many other stresses do not reduce the availability of resources. For example, the effects of temperature extremes or pollutants may be independent of the availability of food resources. These environmental factors will lead to selection for increased stress resistance, and will also increase the instability of population size and thereby decrease competitive interactions.

These considerations suggest that the importance of competition will depend on the types of organisms and environments that are studied. One of the problems in discussions on competition, pointed out by Strong (1983) and others, is that much of orthodox competition theory refers to birds, lizards, and other vertebrates which tend to undergo only small fluctuations in population size and are only a small component of biotic diversity. In contrast, Strong considers the decomposer insects, which comprise a large component of the biota and live in habitats consisting of highly dissected, rapidly changing resource patches rather than homogeneous patches at an equilibrium, and writes that:

Interspecific competition is probably not commonly important for herbivorous insects because autecology, vertical food-web factors, and the weather normally serve to maintain populations below densities that would deplete resources.

Competition may also interact with stress by influencing levels of stress resistance. In Chapter 1, the point was made that any deviation from optimal conditions is likely to result in a metabolic cost. Exposure to elevated temperatures and other stresses leads to a cost which in turn reduces energy for activities such as obtaining food and reproduction. Eventually, environments may become so extreme that little more than maintenance is possible. Under these circumstances factors such as disease, predation, and competition could be effective in crossing the threshold to lethality because they impose an additional metabolic cost. Competition may therefore decrease the stress resistance of organisms by narrowing tolerance limits. An example is given by Peterson and Black (1988) who examined two congeneric species of suspension-feeding venerid bivalves, *Katelysia scalarina* and *K. rhytiphora*. They found that a history of crowding increased an organism's susceptibility to the stress of sedimentation, so that a combination of biotic and abiotic factors can push the physiological state of an organism closer to lethality.

The above discussion is concerned with the community level and does not consider the possibility of interactions between competitive ability and stress resistance at the genetic level, such as whether or not traits associated with increased competitive ability are associated with increased or decreased stress resistance. We will consider such interactions in later chapters.

2.5 Summary

In the fossil record, the taxonomic as well as the ecological groupings of Phanerozoic genera show the same extinction profiles. This means that extinction is physically driven, and temperature changes have been implicated.

Short periods of extreme climatic stress that cause heavy mortality occur sporadically and unpredictably. These are the periods when population extinctions and major adaptive shifts within species are most likely.

Extremes of climatic variables are closely associated with the margins of many species and may therefore determine species distributions. This has been validated in some cases by the development of predictive bioclimatic envelopes in explaining the distribution of various taxa on the Australian continent and by matching the resistance of species in laboratory experiments to their distributions.

Environmental stress forms a major component of models that attempt to link habitats with species characteristics and community interactions. These general models suggest that stress has played a major role in the evolution of life histories and growth patterns in plants and animals.

Periods of environmental stress decrease the importance of competition in natural populations by contributing to community instability. Stressful conditions may also enhance the intensity of competition over short periods by decreasing the availability of resources. Competition can enhance the effects of environmental stress in some circumstances.

3. Stress and protein variation

We start an overview of genetic variation in stress response at the protein level, before moving to the chromosomal, developmental, and organismic levels in Chapters 4 and 5. Here we consider effects of environmental stress on enzymes and heat-shock proteins, and conclude by discussing the importance of stress in selection on protein polymorphisms.

3.1 Temperature adaptation of enzymes: interspecific and interpopulation variation

Temperature influences the synthesis, stability and activity of enzymes. Temperature changes frequently have substantial effects upon the equilibrium constants of biochemical reactions, especially those involving the reversible formation of noncovalent (or weak) chemical bonds. All of the diverse biological structures stabilized by weak chemical bonds share a common property of changing during the performance of their activities, i.e. they are not rigid and invariant. A fine balance between lability and stability occurs. Hochachka and Somero (1984) write:

The reliance on weak chemical bonds dictates a sharp temperature dependence of macromolecular structure, and this dependence in turn makes the metabolic apparatus and the regulation thereof highly sensitive to temperature change. Likewise the rate effects of temperature will impact on the velocities with which metabolic flux occurs.

In fact, in a consideration of critical structural and kinetic properties of enzymes, Somero (1978) concludes that the main trait not sensitive to biological temperature change is primary structure. It is therefore expected that natural selection will have produced altered enzymes in response to changes in environmental temperature so that they can maintain normal function as much as possible. This will apply particularly to enzymes involved in central metabolic pathways common to all organisms. Kinetic parameters for enzymes involved in peripheral pathways tend to be less conserved, since their importance will depend on the intermittent presence of environmental substrates experienced by an organism.

The conservation of the kinetic properties of enzymes isolated from different biological systems has been demonstrated in a number of studies comparing the temperature responses of homologous proteins isolated from organisms adapted to different environments (Hochachka and Somero 1984; Somero 1986). For example, in the skeletal muscle isozyme of the glycolytic enzyme lactate dehydrogenase (M4-LDH), the Michaelis–Menten constant (K_m) of a substrate (pyruvate) was strongly conserved for a wide range of organisms including a mammal (rabbit), a reptile (the desert iguana, *Dipsosauris dorsalis*) and a variety of fishes when K_m values were measured at the environmental temperature experienced by these organisms. The organisms covered a range of body temperatures from −1.86°C (Antarctic teleost fishes) to 47°C (the desert iguana). The K_m value represents the substrate concentration at which the initial reaction velocity or flux is half the maximal velocity and its conservation indicates that enzymes from different organisms show a similar efficiency in converting this substrate of intermediary metabolism.

The study of closely related species in habitats differing in temperature offers an experimental approach for investigating environmental thresholds of protein perturbation and the types of amino acid substitutions that are effective in maintaining optimal protein properties. As an example, Graves and Somero (1982) describe how barracuda species (genus *Sphyraena*) of the eastern Pacific are found in cool temperate zones on both sides of the equator, in slightly warmer waters of the Gulf of California, and in warm tropical waters. These habitats differ by 6–8°C (Table 3.1). The *Sphyraena* congeners are pelagic schooling fishes with similar body forms and general ecologies. Differences in the kinetic properties of homologous enzymes from these species are therefore likely to reflect adaptations to the thermal environments that these

Table 3.1. Kinetic parameters for the lactate dehydrogenase reactions of three barracuda congeners: *Sphyraena argentea* (temperate), *S. lucasana* (subtropical), *S. ensis* (tropical). (After Graves and Somero 1982.)

	S. argentea	*S. lucasana*	*S. ensis*
K_m of pyruvate at 25°C	0.34 ± 0.03 mM	0.26 ± 0.02 mM	0.20 ± 0.02 mM
Temperature midrange (TM)	18°C	23°C	26°C
K_m of pyruvate at TM	0.24 mM	0.24 mM	0.23 mM

organisms experience, although it is always difficult to completely rule out the possibility of other reasons for any differences between the species. The K_m values of M4-LDH for pyruvate differed at 25°C. However, when at the temperature mid-range experienced by the species in their natural habitats K_m values are virtually identical (Table 3.1), in accordance with the similar comparison of unrelated species for this enzyme described above. This suggests that different thermal environments have selected for enzymes that maintain central metabolite concentrations constant, irrespective of the habitat temperature. Differences in average body temperature of only 3°C were therefore sufficient to favour alternative forms of M4-LDH (Graves and Somero 1982) which may vary by no more than a single amino acid replacement (Somero 1986). Hence responses to new thermal environments may be accomplished by minimal changes in protein structure, at least when temperature changes are not too large.

Enzyme differences between species have also been found at a more localized level by comparing related species occupying somewhat different thermal niches in the same area. Liu *et al.* (1978) examined the thermal sensitivities of malate dehydrogenase (MDH) from two species of cattails (genus *Typha*) growing in the same lake. The lake received heated water from a nuclear power station. *T. domingensis* was not found in areas where maximum water temperatures exceeded 30°C, while *T. latifolia* occurred at higher temperatures. This ecological difference was matched by the increased thermostability of MDH from *T. latifolia* at higher temperatures. The six major MDH isozymes from *T. domingensis* were denatured at 50°C, while only half the isozymes from *T. latifolia* were denatured at this temperature.

Differences in enzyme function at the population level have been described in several widespread plant species. McNaughton (1974) studied NAD malate dehydrogenase from cloned samples of *Typha latifolia* from a range of climates. *T. latifolia* has one of the broadest climatic ranges known among organisms, spanning an area from the Arctic circle to the Equator. McNaughton used samples that had been grown in cultivation for periods ranging from 3 to 7 years to obviate the possibility of long-term acclimation to native habitats, and characterized the activation energy (E_a), thermostability, and activity levels of enzymes from different clones.

E_a is a measure of an enzyme's ability to overcome thermal thresholds in a reaction. Increasing the reaction temperature for a given enzyme decreases the activation energy required. This means that plants from warmer climates will have a similar E_a to those from colder climates when tested in their respective habitats if plants from warmer climates have enzymes with a higher E_a than than those from colder climates when they are tested at the same temperature. In accordance with this, McNaughton found that E_a was positively correlated with the mean temperature of the hottest month. The thermostability of enzymes was also found to be positively correlated with mean summer temperature. The enzymes of organisms from warmer climates were therefore more stable at higher temperatures. Enzyme activity was associated with the frost-free period rather than summer temperatures. When populations were clustered according to these three enzyme properties, three distinct clusters emerged (Fig. 3.1):

(1) populations from long-growing season, maritime sites;

(2) populations from summer hot, continental sites;

(3) populations from short growing season, montane sites.

NAD malate dehydrogenase has therefore evolved thermal properties permitting adaptation to a wide range of climatic habitats. It is noteworthy that the environment is expressed in terms of extremes of hot and cold, providing parallels with studies at the organismic and biogeographic levels (Chapter 2).

Other plant studies have incorporated different environmental conditions so that acclimation responses as well as genetic variation between

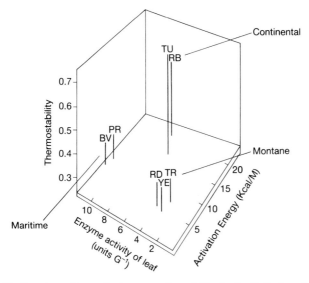

Fig. 3.1. Ordination of population samples of *Typha latifolia* in relation to the activation energy, thermostability and activity of NAD malate dehydrogenase: PR = Point Reyes, California; BV = Beaverton, Oregon; RB = Red Bluff, California; TU = Tully, New York; RD = Redmond, Oregon; Tr = Truckee, California; YE = Yellowstone, Wyoming. (Modified from McNaughton 1974.)

localities can be considered. This approach was used to characterize
NAD malate dehydrogenase from eight *Lathyrus japonicus* clonal
populations collected over a 16° latitudinal gradient in eastern North
America (Simon 1979). The activation energy increased dramatically in
most populations when plants were acclimated under warmer growth
conditions (Fig. 3.2), while thermostabilities were not greatly affected.
Differences between the clones were also evident. Clones from warmer
summer sites had enzymes with higher activation energies (Fig. 3.2) and
enhanced thermostabilities compared to those from colder sites. There
were no differences in enzyme activity between clones when plants were
acclimated at the lower temperatures, but clones from warmer sites had
higher enzyme activities at the warmer acclimation temperatures than
those from colder sites. These clonal differences were similar to the *T.
latifolia* results of McNaughton (1974). Kinetic properties of enzymes
appear to be altered through evolutionary time to fit a particular thermal
environment, and climatic extremes may be important in the selection
process.

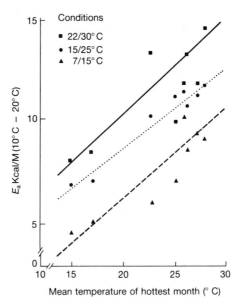

Fig. 3.2. Relationship of the apparent energy of activation of malate dehydro-
genase to summer temperature at collection sites for eight genotypes of
Lathyrus japonicus grown under three experimentally-induced thermoperiods.
Best fit lines are given as a function of the mean temperature of the hottest
month for plants grown at 7/15°C, 15/25°C, and 22/30°C thermoperiods.
(Adapted from Simon *et al.* 1986.)

As a final example, Simon *et al.* (1986) examined the kinetic properties of NAD malate dehydrogenase from clones of three *Viola* species collected over the latitudinal range of each species in eastern North America. Plants were acclimated under three thermoperiods: 7/15°C, 15/22°C and 22/28°C. There were large acclimation effects on the Michaelis–Menten constant (K_m) for oxaloacetic acid, the K_m values tending to increase with the acclimation temperature. There were also differences in the acclimation responses of the clones and species that could be interpreted in an adaptive manner. K_m tended to increase more rapidly with increasing temperature in clones from colder regions than in those from warmer regions, and this led to a lower K_m for clones from warmer regions when plants were acclimated to warmer temperatures. The K_m of enzymes from plants from different regions therefore tended to be more similar at their normal growth temperatures, in agreement with the conservation of K_m values at the interspecific level described for *Arabidopsis thaliana* (Simon *et al.* 1983). This suggests that selection has led to the conservation of kinetic properties for this enzyme to keep the concentration of metabolites of intermediary metabolism constant.

These findings suggest adaptive differences in enzyme properties that reflect climatic selection on populations and species. However, the evidence that differences in kinetic properties contribute to variation in performance under different environmental conditions is indirect. Enzyme variation has not been associated with variation in fitness at the phenotypic level. This is an important limitation of population studies because differences between populations may reflect factors other than kinetic parameters of the specific enzymes under study. Populations may have diverged genetically for a large number of enzyme loci and other loci contributing to performance variation in a range of environments. Testing the contribution of an individual enzyme would require the transfer of genes controlling enzyme kinetics from one population to the genetic background of another population so that the effects of the enzyme can be studied independent of these other genetic differences between populations.

The research on individual enzymes isolated from tissues also ignores other aspects of temperature adaptation at the biochemical level. Somero (1978) considers that a complete understanding of temperature adaptation can only be achieved if one can encompass the enzymes, the solutions which bathe them, and membrane lipids, together with interactions between these components. It is important to replicate *in vitro* the microenvironmental conditions encountered by enzymes *in situ*. Furthermore, enzymes need to be studied in situations that are extreme if responses to environmental stresses are to be understood at the protein level.

An important aspect of temperature–protein adaptations is that a protein must exist in a 'semi-stable' state if it is to have the ability to undergo the changes in shape that catalysis and regulation require (Hochachka and Somero 1984). Membranes must be in a similar state in order to maintain function; this represents a state between extreme fluidity and a rigid gel. The maintenance of this liquid crystalline state is known as 'homoeoviscous adaptation', and there is an extensive litera-ture on the effects of stressors on state changes (e.g. Quinn 1989). As for adaptations in enzyme structure and functional properties, adaptations in lipid systems for responses to temperature perturbations appear ubiquitous among organisms (Hochachka and Somero 1984). Acclim-ation to high or low temperatures involves changes in the percentage of unsaturated lipids to maintain the homoeoviscous state. At low tem-peratures, membranes are more likely to enter the gel phase. Incor-porating unsaturated lipids increases the double bond content and lowers the transition temperature between the fluid and gel phases. Incorporating saturated lipids has the opposite effect, and helps to maintain the homoeoviscous state at high temperatures. This mechanism occurs in bacteria, plants, and animals (White and Somero 1982).

There is evidence that lipid differences and other membrane proper-ties are associated with interspecific variation in resistance to extreme temperatures. For example, Murata and Yamaya (1984) compared major lipid classes in chilling-resistant and chilling-sensitive plant species and found that one lipid class (the phosphatidylglycerols) passed from the liquid crystalline phase to the phase separation phase at a lower temperature in chilling-resistant species. Similarly, Raison et al. (1979) found a correlation between the geographic ranges of plant species and the temperature at which the cell membrane changes from the liquid crystalline state to the gel state. Additional evidence for an association between membrane phase transitions and low temperature is discussed in Woodward (1987).

Membrane lipids may also be important in resistance to other stresses that cause membrane damage. For example, ethanol damages mem-branes in *Drosophila*, yeasts, and other organisms, and Geer and Heinstra (1990) showed that increased ethanol resistance in *D. mel-anogaster* may involve changes in membrane lipids. They examined the association between larval ethanol resistance and physiological traits in seven isochromosomal lines of *D. melanogaster*. There were large positive correlations between survival to pupation and per cent un-saturated fatty acids in total lipid ($r = 0.96$) and between survival to pupation and the percentage of 18-carbon fatty acids ($r = 0.75$), indi-cating that ethanol resistance was increased by unsaturated fatty acids and 18-carbon fatty acids. These types of fatty acids are abundant in *D.*

melanogaster membrane lipids, suggesting an association between genetic variation for ethanol resistance and membrane lipids.

In summary, studies of enzymes at the species and geographic levels indicate that kinetic parameters can often be associated with the climate experienced by organisms, in that enzymes appear to function with greatest efficiency or regulatory ability under the conditions an organism normally encounters. Relatively small temperature differences between habitats may be associated with enzyme differences at the interspecific and intraspecific levels. The modification of kinetic parameters by acclimation can also be interpreted in an adaptive manner. However, the contribution of enzyme variation to fitness differences between populations is not known because populations may differ for a large number of enzymes and other traits. Evidence that enzyme variation can be related to fitness differences within populations under stressful conditions will be discussed below.

3.2 Stress-induced proteins

Cultured cells or whole organisms respond to elevated temperatures by synthesizing a small number of proteins known as the heat-shock proteins or stress proteins, and the synthesis of most other cellular proteins is simultaneously switched off. For example, in *Drosophila melanogaster* flies, normally grown at 25°C, heat-shock proteins are induced when the temperature is raised to 29–38°C, giving a maximum response at 36–37°C within 4 minutes (Schlesinger *et al.* 1982). Simultaneously, the transcription of previously active genes and the translation of pre-existing messages is repressed. So long as cells are maintained at high temperatures, heat-shock proteins continue to be the primary products of protein synthesis. When cells are returned to normal temperatures, normal protein synthesis gradually resumes. In other organisms, the induction of heat-shock proteins is equally rapid, but the maximum induction temperature varies, since it is correlated with the normal range of temperatures which the organisms encounter in the environment (Lindquist 1986). In organisms that grow over a broad range of temperatures, the maximum response is usually achieved at 10–15°C above the optimum growth temperature, while in organisms growing in a more restricted range, maximum response occurs at about 5°C above the optimum.

This response is universal, and has been observed in every organism in which it has been sought, from bacteria to higher plants and vertebrates, suggesting that heat-shock proteins have a very general protective function against elevated temperatures (Lindquist 1986; Lindquist and

Craig 1988). Work on *Drosophila* heat-shock proteins has been extensive, and their genes were among the first eukaryotic genes to be cloned, have their organization within the genome defined, and have their regulatory sequences and associated transcription factors characterized. One of the striking findings is the high degree of conservation of heat-shock proteins throughout evolution, in both protein-coding sequences and regulatory sequences. For example, the complete amino acid sequence of the human heat-shock protein hsp 70 is 73 per cent identical to the *Drosophila* protein, and 50 per cent identical to the analogous *Escherichia coli* product.

It is usually assumed that the heat-shock response protects organisms from the disruptive effects of heat. Supporting evidence has been obtained from acclimation experiments where a group of cells or organisms is split into two identical samples. One sample is shifted directly to an extreme temperature, while the other is incubated at a more moderate elevated temperature, in order to induce heat-shock proteins, before being exposed to the same extreme temperature as the first sample. In the second sample, the normal expectation is a dramatic increase in survival, often by two orders of magnitude or more (Lindquist 1986), reflecting a substantial increase in resistance to heat. This result has now been replicated with a broad variety of organisms at various developmental stages, and in cultured cells under many different conditions.

While this correlation is strong, there is still debate over the direct causal effect of heat-shock proteins on the development of resistance to heat. Lindquist (1986) presents several lines of evidence in favour, and while noting some contradictions in the literature concludes that:

overall, a great many experiments support the notion that heat-shock proteins are crucial to the induction of thermotolerance.

Supporting evidence includes:

(1) treatments that block the induction of heat-shock proteins which abolish the increase in resistance in cells;
(2) the isolation of mutants with blocked hsp induction that do not show resistance.

In addition, the kinetics of resistance correlates well with heat-shock protein synthesis (Schlesinger *et al.* 1982; Marx 1983), a relationship that has been particularly well demonstrated in experiments on mammalian cells (Li and Laszlo 1985). Some difficulties in interpretation probably derive from the complexity of the physiological effects of heat. However, it is also apparent that not all proteins synthesized during the response are involved in heat resistance, since mutants for major heat-shock proteins may not have altered levels of heat resistance (e.g. Ramsay 1988).

Most work on the involvement of heat-shock proteins in acclimation responses has been carried out on cultured cells or unicellular organisms. Not all cases of induced high temperature resistance in multicellular organisms are associated with heat-shock proteins. For example, the conditions for the induction of heat-shock proteins and thermal resistance do not coincide in salamanders, suggesting that they are independent processes (Easton *et al.* 1987). These proteins may be a component of acclimation responses by cells, but additional forms of acclimation involving the neuromuscular system may occur in complex multicellular organisms. Research on heat-shock proteins has also been largely confined to the laboratory, and it is only recently that induction of these proteins has been demonstrated in the field (Spotila *et al.* 1989).

It has become clear that many heat-shock proteins are stress proteins generally, rather than specific responses to heat stress. They are induced by a variety of stressors, including heavy metals, ethanol, sulphydryl reagents, amino acid analogues, virus infections, and oxygen deprivation (Table 3.2: Schlesinger *et al.* 1982; Marx 1983). Moreover, low as well as high temperatures may induce these proteins (Burton *et al.* 1988). It

Table 3.2. Agents or treatments that activate heat-shock protein genes. (Modified from Ananthan *et al.* 1986.)

Inducing agent or treatment	Proposed effects
Group I	
Ethanol	Translation errors
Amino acid analogs, puromycin	Abnormal proteins
Group II	
Heat shock	Increased unfolding of proteins
Cold shock	?
Various heavy metals, copper-chelating agents, arsenite, iodacetamide, p-chloromercuribenzoate	Binding to sulphydryl groups, conformational changes in proteins
Anoxia, hydrogen peroxide, superoxide ions and other free radicals	Oxygen toxicity, free radical fragmentation of proteins
Ammonium chloride	Inhibition of proteolysis
Amytal, antimycin, azide, dinitrophenol, rotenone, heptyl-hydroxy-quinoline N-oxide, ionophores	Inhibition of oxidative phosphorylation, changes in redox state, covalent modifications of proteins
Hydroxylamine	Cleavage of asparagine–glycine bonds in protein

therefore appears best to refer to a large number of the genes as stress-induced genes, and to describe the entire phenomenon as a generalized stress response (Ananthan *et al.* 1986).

Few experiments have considered the importance of stress proteins in increasing resistance to stresses other than heat, although heat shock has been associated with increased resistance to other stresses. For example, in *Neurospora crassa,* heat shock induces high levels of peroxidase which can lead to resistance against normally lethal levels of hydrogen peroxide (Kapoor and Lewis 1987). In cells of the bacterium, *Zymomonas mobilis,* heat-shock proteins are synthesized in response to sub-lethal levels of ethanol and heat, and these treatments increase resistance to both stresses (Michel and Starka 1987). The way in which heat-shock proteins might increase such resistance is unclear, although some may protect proteins from denaturation, or repair partly denatured proteins, using ATP to disrupt aggregates formed by partly denatured proteins and releasing proteins to reassemble to their correct structure (Anathan *et al.* 1986; Pelham 1986). These mechanisms could provide a general form of resistance to stresses that lead to protein denaturation.

While much of the emphasis in heat-shock protein work has been on their expression in response to environmental variables, these proteins may also contribute to genetic variation in stress resistance within populations. There is evidence for the existence of polymorphism in heat-shock proteins among natural populations of *Drosophila* (Petersen *et al.* 1979), and Stephanou *et al.* (1983) have linked heat-shock protein variation with heat resistance in *D. melanogaster.* Stephanou *et al.* (1983) successfully selected lines that were sensitive or resistant to a heat shock and found that ovaries and salivary glands from resistant flies incorporated more labelled methionine after heat shock than the sensitive line (Fig. 3.3), indicating a higher rate of production of heat-shock proteins. This was confirmed by the intensity of labelling of heat-shock proteins on acrylamide gels. In a different experiment, Alahiotis and Stephanou (1982) compared two population cages, one kept at 25°C and the other at 18°C or 14°C for 7 years. The population kept at the higher temperature was relatively more resistant to a heat shock and showed more protein synthesis following a temperature shock, suggesting quantitative differences in the production of heat-shock proteins. A relationship between heat-shock protein variation and thermal sensitivity in wheat has also been suggested (Zivy 1987).

Genetic variation in heat shock responses may also arise through interactions between stresses. For example, Stephanou and Alahiotis (1986) found that variation at the alcohol dehydrogenase (Adh) locus in *D. melanogaster* influenced heat resistance via an ethanol induction effect. Holding flies on medium containing 2 per cent ethanol had a

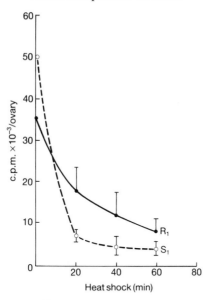

Fig. 3.3. Incorporation of ^{35}S-methionine into proteins of ovaries for different times of heat shock in stocks of *Drosophila melanogaster* that were resistant (R_1) and sensitive (S_1) to heat shock. (Simplified from Stephanou *et al.* 1983.)

protective effect on the survival of adult flies homozygous for the *Adh^F* allele during heat shock of 40°C for 25 minutes. However, ethanol treatment did not induce heat resistance for *Adh^{S}Adh^S* homozygotes. This concentration of ethanol is probably not a stress because it does not decrease viability (McKenzie and Parsons 1972), but ethanol is a stress at higher concentrations. The induction of heat resistance may be related to heat-shock protein synthesis; more protein was synthesized in ovaries from *Adh^{F}Adh^F* flies after a heat shock than in ovaries from *Adh^{S}Adh^S* flies, regardless of whether or not they were pretreated with ethanol.

3.3 Protein polymorphisms and stress resistance

In recent years emphasis has been placed on studies at the protein level in attempts to understand microevolutionary processes. An enormous body of data has accumulated on electrophoretically detectable enzyme variation, and this has been used to make inferences about the relative importance of evolutionary factors influencing genetic changes within populations. Much of the early work on the association between fitness and allozyme variation focused on differences between genotypes under favourable laboratory conditions, but some of the more recent studies have emphasized the performance of genotypes in unfavourable conditions that influence flux through metabolic pathways in which the enzyme is involved.

Allozyme variation has been related to environmental stress in several instances. Populations of the slender wild oat, *Avena barbata*, in the Mediterranean climatic regions of California are often polymorphic at morphological and allozyme loci (Jain 1969; Clegg and Allard 1972). Populations occurring on well-watered, fertile sites are usually monomorphic for a single five-locus allozyme genotype (mesic), while populations on the driest, least-fertile sites are monomorphic for a second genotype (xeric). In the Napa Valley, there are hilly areas where the habitat changes rapidly from deep, rich, mesic soils to poor, shallow, xeric hillside soils. In such a heterogeneous habitat, Hamrick and Allard (1972) demonstrated that frequencies of the mesic and xeric alleles were correlated with environmental changes. Table 3.3 lists *A. barbata* populations from one hillside, ranked by Nei's genetic identity measure according to habitat or origin. The degree of genetic heterogeneity on the hillside is highly correlated with environmental variation expressed in terms of the degree of aridity, in that the genetic identity with a mesic genotype is greatest for mesic populations and becomes reduced as populations come from more xeric habitats. Both at this local level and on a California-wide basis, variation measured from an assessment of electrophoretic variants is only interpretable in terms of differences in adaptation to aridity and temperature regimens which have occurred within the 250 years since *A. barbata* was introduced into the region

Table 3.3. Listing of *Avena barbata* populations from a hillside ranked by Nei's genetic identity measure. Numbers or letters refer to populations. (After Hamrick 1979.)

Identity group				
M	IM	I	IX	X
F (0.992)	B (0.773)	O (0.532)	D (0.361)	6 (0.183)
P (0.991)	4 (0.617)	H (0.525)	K (0.343)	3 (0.177)
1 (0.914)		G (0.518)	I (0.208)	2 (0.157)
C (0.899)		J (0.490)		A (0.056)
E (0.851)				7 (0.038)
N (0.818)				5 (0.036)
				M (0.022)
				L (0.015)

M, mesic; IM, intermediate mesic; I, intermediate; IX, intermediate xeric; X, xeric. Genetic identities, determined by comparing each population with the mesic genotype (21112), are given in parentheses.

(Clegg and Allard 1972). These results clearly demonstrate the occurrence of intense selection involving stress from the physical features of habitats, to the extent that local races are now characterized by particular electrophoretic genotypes. Temperature and moisture-related variables are also significantly correlated with particular isozyme phenotypes in Israel (Kahler *et al.* 1980). Comparisons of the patterns and extent of genetic variation in Israel and California led to the conclusion that the evolution of 'ecotypes', each adapted to a specific habitat and marked by a particular set of enzyme alleles, has proceeded further in Israel, where *A. barbata* is endemic, than in California, where it is a recent introduction. In this example, allozyme morphs have been related to habitat type but the relationship between enzyme variation and selective factors is not understood.

A few attempts have been made to relate enzyme polymorphisms directly to stress resistance. In *Drosophila melanogaster*, two common alleles (Adh^F, Adh^S) can be distinguished at the alcohol dehydrogenase (Adh) locus and numerous studies have attempted to relate allelic variation to environmental stress (reviewed in Chambers 1988). Early work concentrated on environmental ethanol which is a major substrate for Adh. Results were conflicting: in some tests $Adh^F Adh^F$ homozygotes were less resistant to ethanol vapour than the other genotypes (Oakeshott *et al.* 1980; Ziolo and Parsons 1982); however, the reverse was found in other tests where flies or immatures were exposed to media containing ethanol (e.g. Kamping and Van Delden 1978; Oakeshott *et al.* 1980). It has proved difficult to relate allele frequencies to environmental ethanol, even though parallel clines on different continents suggest spatial variation in selective factors (Oakeshott *et al.* 1985). One possibility is that ethanol resistance may be conferred by combinations of genotypes from the *Adh* locus and other polymorphic enzyme loci (McKechnie and Geer 1988).

Other studies, such as the one by Stephanou and Alahiotis (1986) discussed above, have focused on the association between *Adh* genotypes and temperature at the whole organism and enzyme levels. Alahiotis (1982) and others have found that the enzyme produced by the Adh^F allele is less thermostable than that of the Adh^S allele and has a higher K_m for ethanol at high temperatures. The frequency of the Adh^F allele decreased less rapidly in cages kept at 18°C than in cages kept at a warmer temperature (25°C). At the interspecific level, the kinetic properties of alcohol dehydrogenase (K_m for ethanol and Q_{10}) show an association with geographical origin (and hence climate) rather than with phylogeny (Alahiotis 1982). The K_m of the enzymes from tropical species decreased with assay temperature and the K_m of that from temperate species increased, while Q_{10} values for tropical species were higher at high temperatures and the reverse was found for temperate species. These results suggest temperature-associated selection of alcohol dehydrogenase, although only two tropical and temperate species were compared and more species of each group need to be tested.

The alcohol dehydrogenase locus has also been studied in the soft bromegrass *Bromus mollis* (Brown *et al.* 1976). One of the homozygotes ($Adh^S Adh^S$) showed lower catalytic activity and produced 13 per cent more dry matter under continuous flooding whereas there were no differences between genotypes under control conditions. The other homozygote ($Adh^F Adh^F$) germinated more rapidly at 2°C but not at 15°C. Differences between genotypes were therefore apparent only under stressful conditions of low temperature and flooding. The lower catalytic activity may be an advantage under anaerobic flooding conditions because toxic ethanol builds up less, while the more rapid regeneration of NAD by alcohol dehydrogenase with higher catalytic activity may be an advantage in seed development. However, these possibilities were not tested further.

A more clear-cut case where enzyme function has been related to environmental stress is in the marine intertidal bivalve, *Mytilus edulis*, in Long Island Sound, where genetic variation at the leucine aminopeptidase (Lap) locus is associated with levels of environmental salinity (Koehn 1978). The Lap^{94} allele shows dominance for high enzyme activity, and genotypes with this enzyme accumulate free amino acids at a greater rate than other genotypes in a hyperosmotic solution (Hilbish and Koehn 1985). In conditions of low salinity, free amino acids and ammonia are excreted and the Lap^{94} allele also shows partial dominance for rate of excretion. This may account for clines in *Lap* alleles at estuaries where the frequency of the Lap^{94} allele decreases from the ocean to areas of low salinity (Hilbish and Koehn 1985). Selection at this locus was demonstrated in experiments with laboratory-reared *M. edulis* by Beaumont *et al.* (1988). The Lap^{94} allele was selected against in cultures kept at low salinity during the post-larval stage.

A related example (Burton and Feldman 1983) can be found in the copepod *Tigriopus californicus*, which regulates cell volume during osmotic stress by accumulating alanine, proline and glycine. There are two common alleles at the structural locus for the enzyme glutamate–pyruvate transaminase, which catalyses the final step of alanine synthesis. One of the homozygous genotypes at this locus accumulated less alanine, and larvae with this genotype were less resistant to hyperosmotic stress than larvae with the other genotypes. Larval mortality differences were fairly large, the mean mortality of the resistant genotypes being 13 per cent compared with 39 per cent for the sensitive homozygote. Genotypic differences are likely to be larger under conditions of nutritional stress, when alanine becomes important in osmo-regulation (Goolish and Burton 1989). Genotypic differences in survival corresponded to variation in specific activity of the allozymes (Burton and Feldman 1983). This example, and the two preceding studies, are concerned with environmental filtering (Watt 1983) where changes in enzyme activity can help to minimize the impact of an external environmental stress on the internal environment.

Smith *et al.* (1983) examined electrophoretic variation at a malate dehydrogenase locus in a population of largemouth bass. One of the alleles (*Mdh-1ᵃ*) was present at a high frequency in fish in a pond receiving thermal effluents from a nuclear power plant. Water from this plant was at over 70°C and fish only existed in peripheral areas of the effluent pond during reactor operation. Fish in another pond, which no longer received effluent, showed a steady decrease in the frequency of this allele to that found in a population living in water mostly at ambient temperature. Genotypes with the *Mdh-1ᵃ* allele were more thermostable because they showed a lower rate of denaturation at 55°C, consistent with different allele frequencies in the ponds.

There are some additional reports of allozyme differences associated with stress where the relationship between the stress and the enzyme function is not known. For example, Nevo *et al.* (1981) demonstrated survival differences between allozyme genotypes at the phosphoglucomutase locus in the shrimp, *Palaemon elegans*, following exposure to mercury: one of the heterozygotes increased in frequency from 23 per cent to 39 per cent after exposure to the stress. The more resistant genotypes were more common in a mercury polluted site compared to several unpolluted sites in the Mediterranean (Nevo *et al.* 1984*a*), but a metabolic or filtering role of this enzyme was not established. Other examples involving pollutants are discussed in Nevo (1988).

More generally, allele frequencies of electrophoretic polymorphisms tend to be correlated directly or indirectly with geographical variation in temperature. There are many examples, including the majority of enzyme or protein polymorphisms in our own species (Piazza *et al.* 1981). Such associations reflect the importance of temperature in all biological processes, and are quite usual if adequate data are collected. Using published data on natural populations of 1111 species from a wide range of geographic locations, Nevo *et al.* (1984*b*) analysed correlations of abiotic and biotic factors with genetic diversity assessed electrophoretically, and concluded that the 20 per cent of the genetic variance that could be explained was mainly associated with ecological heterogeneity. Therefore a major determinant of evolutionary change based upon protein polymorphisms is ecological, and temperature with an emphasis upon extremes, emerged as one of the major variables.

Interpreting variation at the molecular level will undoubtedly become progressively more difficult as the increasing complexity of diversity at the DNA level is unravelled by techniques such as restriction enzyme analysis and the study of transposable elements at the population level. As studies at the molecular level become more sophisticated, it becomes increasingly more difficult to demonstrate selection on molecular variants. Selection can be demonstrated more easily at the phenotypic level in terms of phenotypes relevant to the ecology of an organism, such as traits involved in stress resistance and stress evasion. Underlying such traits is an integration of the physiological effects of many genes incorporating interactions within and between loci; this means that the effects of selection may be difficult to predict and may vary among populations differing in their genetic backgrounds. However, there are situations where ecological phenotypes are controlled by genes of major effect in which convergence between the phenotypic and genotypic approaches occurs as in the examples discussed above.

3.4 Summary

Kinetic properties of enzymes central to metabolism may be conserved in unrelated animal species when they are tested in their natural thermal conditions, suggesting selection for the maintenance of constant concentrations of central metabolites. The activation energy, thermostability, and the Michaelis–Menten constant of enzymes may differ at the geographic level within plant species when populations are tested under the same conditions. Populations experiencing warm temperature extremes tend to have more thermostable forms and higher activation energies than populations from cooler environments. However, the fitness consequences of these biochemical differences are not known.

Heat-shock proteins are ubiquitous and are activated by various metabolic stresses, including temperature. Their role in stress resistance is unclear, as is their contribution to interspecific and intraspecific genetic variation in stress resistance, although some studies suggest that they may contribute to variation in temperature resistance.

A few studies have demonstrated an association between heritable enzyme variation and stress resistance, particularly where the enzyme is involved in environmental filtering. Analyses of protein polymorphism patterns in natural populations indicate a general dependence upon climatic extremes. However interpreting variation is difficult because the effects of natural selection at the molecular level tend to be indirect and may depend on interactions among loci.

4. Genetic variation in stress response

Organisms respond to environmental stress with behavioural, physiological or morphological adjustments to counter its effects and maintain normal functioning. These adjustments can take the form of stress evasion or phenotypic changes to increase stress resistance (Section 1.3). Evasion reduces contact with the stress; resistance involves one of a complex of physiological or morphological responses that enable organisms to survive and reproduce under stressful conditions.

Differences in the ability of individuals to respond to stress may be induced by environmental variation, reflect underlying genetic variation, or both. The extent to which environmental variation can modify the expression of a genotype at the phenotypic level is known as phenotypic plasticity (Section 1.3). Genotypes may differ in their susceptibility to environmental modification: this raises the possibility of genetic variation for the plasticity of a stress response as well as genetic variation for the response itself. For example, consider a strain of *D. melanogaster* that produces more heat-shock proteins than another strain. These strains may have the same resistance to a high temperature stress if the heat-shock response is not triggered. However, the strain producing more heat-shock proteins will probably be relatively more resistant to a high temperature stress when the environment includes conditions that trigger synthesis of heat-shock proteins, such as exposure to sub-lethal temperatures. This strain will therefore have a higher level of plasticity for high temperature resistance.

In the next two chapters we examine the nature of genetic variation underlying stress responses, extending our analyses from the protein level to the phenotypic level. We start in this chapter by outlining some of the evidence for genetic variation in stress resistance and stress evasion within and between populations, before turning to genetic variation for levels of phenotypic plasticity. This is followed by a brief discussion of conditions likely to favour genetic changes or plastic changes when populations are exposed to an environmental stress. The effects of stress on genetic variation are considered in the next chapter.

There are two approaches for investigating genetic variation for stress responses (Fig. 4.1). One approach starts with the phenotypic variance in stress response traits and partitions the variance into genetic and non-genetic components, as an initial step in characterizing the genetic

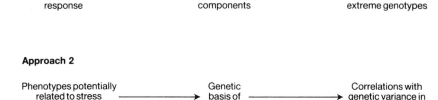

Fig. 4.1. Two approaches for studying genetic variation for stress responses.

variation. This is the standard approach used in quantitative genetics and does not assume any knowledge about the loci contributing to the genetic variance. Nevertheless, strains with extreme values may eventually be isolated, and specific genes accounting for a large proportion of the genetic variance in a stress response trait may ultimately be identified. An alternative approach starts with a genetic polymorphism or variation in a trait with a known genetic basis and attempts to relate this variation to traits involved in stress resistance or evasion. This is essentially the approach used in the study of enzyme polymorphisms discussed in Chapter 3. The importance of the trait or polymorphism under investigation can eventually be determined by examining the amount of genetic variance in a stress response trait accounted for by the polymorphism. For example, quantitative genetic studies suggest that the *Adh* polymorphism in *D. melanogaster* does not account for much of the genetic variance in ethanol resistance (McKenzie and Parsons 1974*a*; Cohan and Hoffmann 1986).

4.1 Stress resistance: pattern and colour variation

We first consider the case of pattern and colour variation as an example where specific traits have been related to quantitative genetic variation in stress resistance. Pattern and colour variants often reflect relatively simple genetic polymorphisms, or represent quantitative traits that have a high heritability so that there is a direct relationship between genotype and phenotype.

Classic examples of this type of variation include the colour and banding polymorphisms of the terrestrial snail *Cepaea*. The frequencies of shell colour morphs in both *C. nemoralis* and *C. hortensis* are associated with climatic variables (temperature, shading, humidity), particularly over large areas that include a range of climatic conditions (Jones *et al.* 1977). Shell morphs differ in their absorption of solar radiation. Detailed field measurements of shell temperatures indicate that dark pigmented shells are almost always hotter than light shells (Heath 1975), suggesting that they will be less resistant to high temperatures. Temperature extremes appear to be particularly important in habitats representing the distribution margins of these species (Arnason and Grant 1976; Bantock and Price 1975); in particular, the brown morph of *C. nemoralis* is abundant in sites with low night temperatures, while mid-banded forms are common in open habitats exposed to temperature extremes. Visual selection via predation appears unimportant at the distribution margins, although this factor affects morph frequencies in more benign habitats.

Richardson (1974) found field evidence for climatic selection within a population of *Cepaea* in sand dunes by comparing morph frequencies in the overall population with morph frequencies among snails that had apparently died from high temperature stress. Numbers of unbanded and yellow forms found among the dead snails were lower than expected (Table 4.1), suggesting that these morphs were more resistant to high temperatures than pink and banded morphs. Direct evidence of temperature-related selection on morphs has also been obtained from adult mortality in field cages (Bantock 1980; Tilling 1983).

C. nemoralis and *C. hortensis* show geographic variation in body colour as well as shell colour, and this has also been associated with temperature: the paler morphs tend to occur where the mean daily

Table 4.1. Distribution of shell morphs in dead *Cepaea nemoralis* collected from sand dunes. Expected numbers based on random sampling of live snails are given in brackets. (Modified from Richardson 1974.)

Banding pattern	Shell colour			
	Yellow	Pink	Brown	Total
Site 1				
00000 (unbanded)	0 (0.5)	3 (10.5)	12 (11.3)	15 (22.3)
00300 (1 band)	12 (12.7)	78 (81.1)		90 (93.8)
12345 (5 bands)	10 (17.6)	130 (111.2)		140 (128.8)
Total	22 (30.8)	211 (202.8)	12 (11.3)	245
Site 2				
00000 (unbanded)	5 (15.2)	27 (45.6)	1 (0.0)	33 (60.8)
00300 (1 band)	10 (21.3)	291 (258.1)		301 (279.4)
12345 (5 bands)	4 (21.3)	169 (145.5)		173 (166.8)
Total	19 (57.8)	487 (449.2)	1 (0.0)	507

maximum temperatures are higher (Cowie and Jones 1985). As in the case of shell colour, darker morphs tend to heat up more quickly, and may be favoured in cooler places because they are able to absorb radiation, while paler morphs may be favoured in hotter places because they can reflect radiant energy giving increased resistance to high temperatures.

As expected, colour and pattern polymorphisms in other snail species also show correlations with physical features of the environment. For example, variations in the shell colour of the intertidal snail, *Nucella lapillus*, have been associated with stressful habitats (Etter 1988). Brown morphs tend to occur on exposed shores while white morphs predominate on protected shores where high temperature and desiccation stresses are relatively more intense. Field and laboratory experiments indicated that brown morphs suffered higher mortality than white morphs in the protected environment because they heated up faster and desiccated more rapidly, although a genetic basis for these differences was not established.

Pattern polymorphisms have been widely investigated in butterflies, particularly in *Colias* where there is an association between temperature resistance and the amount of pigmentation in wing scales. *Colias* from cold environments have darkened underwings whereas those from warm climates do not; this is associated with thermoregulation (Watt 1968; Kingsolver and Watt 1983). Butterflies orientate themselves to increase body temperature and achieve flight at low temperatures; dark morphs heat up faster than light morphs and fly more readily at low temperatures. However butterflies cease flight at high body temperatures (40–42°C) and orientate themselves parallel to solar radiation to avoid overheating because heat shocks cause mortality and decrease fecundity. Light morphs cool more quickly and are therefore more resistant to heat stress. Flight activity is related to fitness, so butterflies are expected to maximize their flight time without becoming heat stressed.

Kingsolver and Watt (1983) studied thermoregulation in three *Colias* populations from different altitudes. The mid- and high-altitude populations had higher pigmentation levels, but showed significantly less flight activity compared to the low-altitude population. Kingsolver and Watt showed that *Colias* from a higher altitude were more likely to experience a heat stress because of increased variability in wind speed and air temperature, despite a lower mean temperature. Thus while higher pigmentation levels helped to increase body temperature, this contributed to the increased susceptibility of the high-altitude population to temperature extremes. As a consequence, high-altitude populations reduced flight activity to keep their mean body temperatures 2–3°C lower than that of low-altitude butterflies, so that they experienced the same maximum temperature. This emphasizes the importance of temperature extremes rather than mean temperature differences between habitats in determining behavioural responses to environmental stress. Wing pigmentation in *Colias* provides an example of genetic variation in a pattern polymorphism mediating a behavioural response to temperature stress. The behavioural response represents a form of stress evasion, even though morphs themselves differ in terms of resistance. Associations between genetic variation in body colour and environmental stress can therefore be complex, and are not necessarily fully understood from isolated experiments on stress resistance. A similarly complex interaction between evasion and temperature resistance may occur in snails: there are behavioural differences in the response to sunlight of different *Cepaea* morphs from the same population (Jones 1982). Thus it cannot always be assumed that different morphs from the same population experience the same stress, or that the climatic differences between sites accurately reflect the levels of stress that populations from these sites are likely to experience. Temperature responses have been associated with variation in pigmentation in many other species, including grasshoppers, locusts, crabs, beetles, aphids and ladybirds (references in Heath 1975).

Colour morphs have also been associated with environmental stress resistance in endotherms. Finch and Western (1977) examined coat colour in indigenous cattle of East Africa and found that lighter coat colour was favoured under conditions of heat stress. Cattle with light coats absorbed less solar radiation and required a lower water intake than cattle with dark coats. However cattle with dark coats appear to have an advantage under cooler conditions because they had higher productivity in areas of low heat stress. Dark colouration in endotherms can reduce the use of energy. For example, white zebra finches dyed black used 22.9 per cent less energy, as measured by oxygen consumption, than non-dyed individuals (Hamilton and Heppner 1967).

In summary, the distribution of colour and pattern variants may often be associated with temperature extremes. Temperature selection may act directly via adult mortality, at least in the case of *Cepaea*, but the stress experienced by morphs may be modified by behavioural evasion. The effects of temperature stress on pattern polymorphisms will probably be more apparent in marginal habitats than in favourable habitats because extreme stresses leading to mortality are more likely to occur at species margins.

4.2 Stress resistance: geographical comparisons

Some of the earliest examples of genetic variation in stress resistance come from genecological investigations of plant species (Section 1.2). In general, reciprocal transplant and common garden experiments indicate that performance differences between plant populations occupying diverse environments tend to have a large genetic component, and populations tend to perform relatively better in the conditions where they originated (reviewed by Heslop-Harrison 1964). These studies have usually concentrated on morphological and life-history traits rather than stress resistance. However differences in resistance are implicated when plants from one habitat fail to survive or show large reductions in growth or reproduction when they are grown in another habitat.

Such differences have been demonstrated for a range of environmental stresses, including light intensity, temperature and soil conditions. For example, Bjorkman (1966) reviewed experiments which compared two populations of *Solidago virgaurea* from shaded habitats on the forest floor to two populations from exposed habitats. When clones were grown in controlled growth chambers under high and low light intensities, plants from shaded habitats grew faster than plants from exposed habitats under low light, but much slower under high light, where plants from exposed habitats attained their maximum growth rate. Mooney and Billings (1961) found that the cotyledonous tundra plant, *Oxyria digyna*, from northern latitudes of North America showed higher survival after exposure to high temperatures than southern populations when grown in a glasshouse. Robson and Jewiss (1968) compared the growth of British and North African varieties of tall fescue (*Festuca arundinicea*) and found that the British type showed less tissue damage and higher survival rates when subjected to a cold temperature stress. Additional cases where the relative performance of populations changed dramatically in different environments include *Festuca ovina* populations from acid and calcareous soils (Snaydon and Bradshaw 1961) and *Dactylis glomerata* populations from different moisture conditions (Ashenden *et al.* 1975).

Associations between climate and genetic variation in stress resistance have also been demonstrated in *Drosophila*. Resistance to climatic stresses in *D. melanogaster* can be measured as the time taken for adults to die after they are exposed to a stress. Stanley and Parsons (1981) found that lethality times varied along the east coast of Australia in a manner predicted by geographical variation in climatic factors. Data on three traits from a population from the temperate zone and from two contrasting tropical climates (humid and wet–dry) are given in Table 4.2. The Melbourne population was more resistant to cold than flies from Darwin, as predicted; Melbourne was also the most resistant to desiccation, followed by Darwin and then Townsville. The relatively high desiccation resistance of Darwin is predicted by the pattern of seasonal rainfall with intervening dry periods in Darwin, while the relatively high

Table 4.2. Mean stress resistance (±95% confidence intervals) of *Drosophila melanogaster* stocks originating from three Australian locations. (Data from Stanley and Parsons 1981.)

	Melbourne (temperate)	Townsville (sub-tropical)	Darwin (wet–dry)
Cold resistance*			
Females	6.6 (± 1.4)	—	10.3 (± 3.3)
Males	19.1 (± 3.4)	—	23.6 (± 0.8)
Desiccation resistance†			
Females	23.2 (± 0.9)	19.9 (± 1.3)	22.1 (± 1.0)
Males	18.5 (± 0.9)	15.4 (± 0.9)	17.8 (± 0.8)
Ethanol resistance‡	57.8 (± 5.7)	7.9 (± 3.9)	34.5 (± 4.4)

*Number of flies dead out of 25 after 40 hours at −1°C.
†Time in hours for 50% of flies to die (LT_{50}) at 0% RH, 25°C.
‡Longevity (LT_{50}) of adults exposed to 12% ethanol in hours.

resistance of Melbourne is consistent with the dry periods experienced at this location. The populations also differed for ethanol resistance although this variation is more difficult to relate to habitat because there appear to be minimal differences in ethanol level between temperate and tropical resources (Gibson *et al.* 1981).

Turning to other species, Timofeef-Ressvosky (1940) studied the relative viability of *D. funebris* from regions covering the major climatic zones of Europe, northern Africa, and into Asiatic Russia at 15°C, 22°C and 29°C. The northern populations were more resistant to cold temperatures, while southern populations were more resistant to high temperatures. Eastern populations showed resistance to both temperature extremes, consistent with the continental climate experienced by the eastern regions where winter temperatures are very low, and summer temperatures are very high. Eckstrand and Richardson (1980) found that a population of the Hawaiian species, *D. mimica*, from a very dry area was more resistant to desiccation than other populations of the same species. Finally, studies in a number of *Drosophila* species have shown that variation in body size leads to variation in stress resistance, and the mean size of populations decreases as the average temperature where populations originated increases (review in David *et al.* 1983).

Geographic variation in stress resistance has been demonstrated in other animal species. Battaglia (1967) summarized results from experiments with populations of the copepod, *Tisbe furcata*, which lives in estuarine and marine environments. The salinity resistance of one population from a marine environment (Plymouth) was compared to that of two populations from brackish water environments consisting of a lagoon (Chioggia) and a lake (Gargano). Stocks had been held in the laboratory for more than 60 generations. Resistance to low salinity was tested by holding adults in standard laboratory sea water and then transferring them to diluted sea water. Mean recovery times after transfer were 81.58 ± 1.86 minutes for Plymouth, 54.42 ± 1.16 minutes for Chioggia and 43.19 ± 0.82 minutes for Gargano. These differences reflected the differences in salinity levels recorded at the locations where the populations originated. As another example, the thermal resistance of clones of the crustacean, *Daphnia pulex*, from different locations correlated with the maximum temperature of the environment from which they originated (Macisaac *et al.* 1985). This correlation was also found in comparisons of *Daphnia* species.

The above findings indicate that geographic variation in stress resistance can be genetically based and often matches climatic predictions. However it should be emphasized that there are some published studies, particularly in plants, where genetic differentiation between populations has not been found. Sultan (1987) has argued that much of the phenotypic variance in plant populations is associated with plasticity, and lists several cases where morphological and physiological differences between populations occupying different habitats have only a minor genetic component. For example, Clough *et al.* (1979) compared photosynthetic properties of *Solanum dulcamara* from sunny and shaded sites and found no evidence of ecotypic differentiation when plants were grown from vegetative cuttings in environments with different light intensities. Individuals originating from the sunny and shaded sites showed similar responses to light primarily involving changes in leaf thickness. There was evidence for genetic variation in photosynthetic properties among individuals from the same site; an absence of genetic variation within populations probably did not account for the absence of interpopulation differences. Rehfeldt (1979) compared families representing 12 populations of *Pinus monticola* by growing plants from seed in two contrasting environments. He found no genetic variation among populations for growth and development, but did find genetic variation among families from the same population. Finally, Antonovics and Primack (1982) made reciprocal transplants with seedlings of a perrenial herb (*Plantago lanceolata*) between six sites, and found that environmental factors contributed to site differences in mortality, growth rate, and fecundity more than genetic factors.

The absence of genetic differences between populations in such studies raises questions as to when populations are expected to respond to environmental stress with genetic changes. One answer is that genetic differences should arise when a single genotype cannot have the highest fitness under all environmental conditions, and different genotypes should therefore be selected under different conditions. This implies that genotypes have different 'norms of reaction' as proposed by Schmalhausen (see Section 1.2). We encountered some examples of this at the enzyme level in Chapter 3. However there are other considerations that may dictate whether population differences are genetic or environmental, and we discuss these below when considering phenotypic plasticity in stress responses.

Geographic differences in stress resistance may not always fit climatic predictions. Coyne *et al.* (1983) examined resistance to heat, cold and desiccation in seven populations of *Drosophila pseudoobscura* from different geographical locations. Significant variation between populations was found in tests with adult flies, but differences were not in the direction predicted by climatic data. In contrast, population differences in the stress resistance of pupae were in the predicted direction. It is difficult to evaluate the significance of these results in the absence of data on stress levels experienced by adults in the field. *D. pseudoobscura* adults may not experience the types of stress levels to which they were exposed in the laboratory, particularly as *D. pseudoobscura* has relatively high levels of stress resistance compared to other *Drosophila* species (Parsons 1982a, 1987a).

The above examples focus largely on climatic stresses. There are also many cases where geographical variation has been demonstrated for specific chemical stresses. These include variation in responses to atmospheric pollutants (Hutchinson 1984) and heavy metals (Antonovics *et al.* 1971) in plants, and marine pollutants in animals (Weis and Weis 1989).

4.3 Resistance variation within populations

Numerous experiments have demonstrated genetic variation in stress resistance within populations of *Drosophila*, plants and other organisms. Much of the early *Drosophila* work in this area, including comparisons of *D. pseudoobscura* inversion strains at extreme temperatures, and experiments involving the characterization of stress resistance in inbred laboratory lines of *D. melanogaster*, was reviewed by Parsons (1974). Attempts to characterize genetic variation in natural populations were usually carried out with the isofemale line approach, which involves the collection of inseminated females from the field and the separate culture of progeny from each female in the laboratory to generate a series of isofemale lines. Comparisons of the variance between lines with the variance within lines can provide an indication of the level of genetic variation in the field population.

The isofemale line approach provided evidence for genetic variation in resistance to many stresses in natural *Drosophila* populations, including high and low temperature, desiccation, CO_2, anoxia, γ-irradiation and various chemical stresses such as ether, chloroform, ethanol, and acetic acid (Parsons 1974, 1980). These studies indicated significant heritabilities for stress resistance traits, although the narrow and broad heritabilities cannot be distinguished from isofemale line data because differences among lines include dominance as well as additive components. Estimates of heritability can be obtained when directional dominance is not strong and each line is kept at a large size to minimize inbreeding (Hoffmann and Parsons 1988). Unfortunately, the size of isofemale lines was not controlled in much of the early work on stress resistance traits, and resistance was measured on groups rather than single individuals, making it difficult to relate variance estimates to heritabilities based on measurements of individuals (Hoffmann and Parsons 1988). These studies therefore provide good evidence for substantial genetic variation in stress resistance when flies are cultured under laboratory conditions, but little information on the nature of this variation.

Genetic variation in stress resistance traits in *Drosophila* has also been demonstrated with artificial selection experiments. Selection for increased resistance to γ-irradiation produced a rapid response, with realized heritabilities in four *D. melanogaster* lines ranging from 0.48 to 0.97, but selection for increased radiation sensitivity was not effective (Westerman and Parsons 1973). We have selected for increased desiccation resistance in three lines of *D. melanogaster*, and obtained realized heritabilities ranging from 0.59 to 0.66 (Hoffmann and Parsons 1989*a*). Selection for increased cold resistance in *D. melanogaster* was successfully carried out by Tucic (1979), despite a low heritability. Selection for increased resistance to a heat shock in *D. melanogaster* was unsuccessful, although increased sensitivity could be readily selected (Morrison and Milkman 1978). Several selection experiments have demonstrated genetic variation for resistance to chemical stresses in *Drosophila*, including insecticides (e.g. Crow 1957; King and Somme 1958) and ethanol (e.g. Gibson *et al.* 1979; Cohan and Hoffmann 1989).

Cohan and Hoffmann (1989) summarized the results of selection for adult resistance to knockdown by ethanol fumes in *D. pseudoobscura* and *D. melanogaster* stocks from five locations, and on *D. persimilis* and *D. simulans* from one location. Significant realized heritabilities were obtained for all the selection lines, but these differed between species and populations. For example, selection on *D. persimilis* gave a heritability of 0.41, while heritabilities for populations of a sibling species (*D. pseudoobscura*) ranged from 0.10 to 0.23. These results indicate that natural populations may have different levels of genetic variation for the same stress resistance trait, and that heritability differences also occur among closely related species.

Heritability estimates for stress response traits have not been extended to field conditions. Field heritability estimates can be obtained in *Drosophila* by measuring the phenotypes of field-collected females and their progeny that have been cultured under laboratory conditions. This approach was used by Coyne and Beecham (1987) to demonstrate heritable variation in body size in *D. melanogaster*. Methods for estimating the standard error of the heritability estimate have recently been derived (Riska *et al.* 1989), and this approach should provide a useful next step in characterizing genetic variation in stress response variation in *Drosophila*, although its validity depends on the assumption that genetic variation expressed under field conditions will also be expressed under laboratory culture. It is possible that variation in the field will not always be expressed in the laboratory (or vice versa) because the parent–offspring correlation may depend on the environmental conditions experienced by different generations (e.g. Van Noordwijk *et al.* 1988). Major shifts in the relative fitness of genotypes can occur during laboratory culture as shown by the process of domestication in *Drosophila* (Kohane and Parsons 1988).

Turning to animals other than *Drosophila*, several studies have examined genetic variation for heat resistance in caged birds because of the commercial importance of this trait in determining survival in hot environments. Using a half-sib design with a random bred population of Japanese quail, Bowen and Washburn (1984) obtained a heritability estimate of 0.32 ± 0.15 from the sire component and somewhat higher values, around 0.5, from the dam component, suggesting non-additive as well as additive genetic variance for heat resistance. Genetic and phenotypic correlations between heat resistance and body weight were negative (the most reliable genetic correlation was given as -0.54), indicating that smaller birds had higher heat resistance. This probably reflects the more efficient heat dissipation by smaller birds because of their greater surface area to volume ratio. Similar results were obtained for white leghorn chickens by Wilson *et al.* (1975). Genetic variation for resistance to temperature extremes within populations has also been demonstrated for some non-domesticated animals, such as wasps (White *et al.* 1970) and copepods (Bradley 1981).

Numerous experiments have demonstrated heritable variation for stress resistance in plant populations. Most of these have been carried out with crop plants grown under a range of climatic conditions, and have involved yield comparisons among varieties or populations derived from combining cultivars (e.g. Johnson and Frey 1967; Abel 1969;

Richards 1978; Rumbaugh *et al.* 1984). Genetic variation has also been characterized by exposing plants to specific stressors. For example, Ashraf *et al.* (1987) described genetic variation in the salt resistance of seedlings in four forage species: forage rape (*Brassica napus*), berseem clover (*Trifolium alexandrinum*), alfalfa (*Medicago sativa*) and red clover (*Trifolium pratense*). They selected within one variety of each species and obtained realized heritabilities in the range 0.31 to 0.62, indicating considerable genetic variation for increased salt resistance. Many other studies with crop plants have revealed high heritabilities for resistance to stresses such as drought, heat, chilling and mineral deficiencies (reviewed by Blum 1988). Genetic variation for resistance to atmospheric pollutants has also been widely documented in crop plants (Ryder 1973). Few experiments so far have considered stress resistance in individuals originating from natural plant populations, although high heritabilities have been obtained for exposure to heavy metals (e.g. Gartside and McNeilly 1974; Bradshaw 1984; Shaw 1988) and air pollutants (e.g. Taylor 1978).

Many discussions on phenotypic variation within natural plant populations have emphasized the importance of environmental factors rather than genetic factors (Harper 1977; Sarukhan *et al.* 1984; Sultan 1987). Because plants cannot move, they are exposed to a wide range of environmental conditions, which makes it likely that immediate responses to environmental stress will involve plastic changes rather than genetic changes (see below). In addition, opportunities for natural selection may be reduced in plants because performance of individuals depends to a large extent on chance (Harper 1977). Few seeds may fall in sites that are suitable for germination, and growth rate subsequent to germination may depend more on the nature of the site than on intrinsic differences between individuals. This means that growth differences between individual plants in natural populations will tend to reflect environmental factors rather than genetic factors.

Nevertheless, associations between microspatial heterogeneity in the environment and genetic markers indicate genetic variation for stress responses in natural plant populations. We have already discussed the example of *Avena barbata* (Section 3.3), and other examples include associations between the resistance of trees to air pollution and electrophoretic markers as found in beeches (Muller-Starck 1985) and spruce (Scholz and Bergmann 1984). Genetic variation for stress responses within natural plant populations has also been shown, as in some of the localized differences in resistance discussed below and in Chapter 6.

In summary, there is abundant evidence for genetic variation of stress resistance traits within populations. However, most experiments with animals have been carried out in the laboratory, and the heritability of stress resistance traits needs to be considered further in the field. Many experiments with plants have been carried out under a range of field conditions, but few experiments have considered genetic variation in natural plant populations.

4.4 Seasonal and microgeographic variation in stress resistance

Genetic variation for stress resistance within natural populations exposed to seasonal stress periods may result in different genotypes being selected at different times of the year, leading to seasonal genetic changes within populations. A few studies have provided evidence for such changes. The crustacean, *Daphnia magna*, reproduces by parthenogenesis so that a population consists of several clones. The frequency of different clones changes markedly between seasons, and Carvalho and Crisp (1987) and Carvalho (1987) related these changes to variation in stress resistance. Strains derived from females representing the most abundant clones in each season were bred under constant laboratory conditions for several generations before their viability and fecundity were scored at a range of temperatures. Clones performed best at temperatures which corresponded to those seasonal temperatures when they were predominant (Fig. 4.2): winter clones performed better at cooler temperatures and summer clones performed better at higher temperatures. The seasonal changes in clone frequencies may therefore be a consequence of genetic differences in stress resistance.

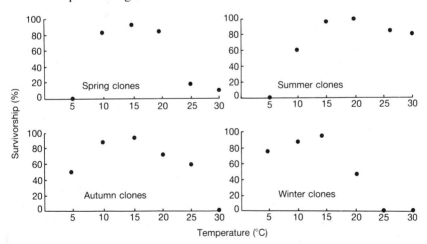

Fig. 4.2. Effect of temperature on the survival of seasonal clones of *Daphnia magna*. Each point represents the mean of four clones. (Simplified from Carvalho 1987.)

Seasonal changes in stress resistance that are genetically based may also occur in sexual species. McKenzie and Parsons (1974b) measured the desiccation resistance of offspring from field-collected females of a Melbourne *Drosophila simulans* population. The offspring were always cultured and tested under the same laboratory conditions. Desiccation resistance was greatest during the warmer months and became lower as the temperature decreased. This pattern was evident over the 3 years of the study and is likely to reflect genetic changes in the population. Seasonal changes in the reproductive potential of *D. melanogaster* females were found in French populations by Bouletreau-Merle *et al.* (1987). Genotypes generating a large number of ovarioles and strong oviposition blocking increased in frequency in spring and autumn, while genotypes producing the opposite phenotypes were favoured in summer. In laboratory populations kept at different temperatures, a large number of ovarioles and strong blocking were favoured at cool temperatures but not at high temperatures, although the reason why different phenotypes were selected at these temperatures is not clear.

Seasonal changes in phenotypes may reflect genetic changes if traits show high heritabilities under field conditions. In house sparrows, post-winter males were smaller than pre-winter males, while the reverse occurred in females (Fleischer and Johnston 1982). These opposing size changes both involved increases in the skeletal core-to-limb ratio because male core size showed a disproportionate increase while female limb size showed a disproportionate decrease. These changes decreased the surface to mass ratio, which can lead to lower metabolic rate and heat loss—features likely to increase resistance to exposure to severe winter weather conditions. A genetic basis for these changes was not demonstrated, although field heritabilities for morphological traits in birds tend to be high (e.g. Boag and Grant 1978; van Noordwijk *et al.* 1988).

From these few examples, it is unclear how common genetic changes in stress resistance over seasons are. The demonstration of seasonal genetic changes requires long-term studies of natural populations and few relevant studies have been published. In *Drosophila*, there are several examples of seasonal changes in inversion polymorphisms (reviewed in Parsons 1974) and in allozyme polymorphisms (e.g. Gionfriddo *et al.* 1979; Nielsen *et al.* 1985), which suggest that genotypes may differ in fitness under different seasonal conditions. Genetic changes over time in response to climatic factors may even occur on a daily basis, as indicated by differences in the frequency of lethal genes at midday as opposed to morning collections of *Drosophila willistoni* in the tropics (Hoenigsberg *et al.* 1977).

Microgeographic variation in stress resistance has been investigated in a few situations. The classic case is heavy metal resistance in plants. Many plant species have populations with the ability to grow on soils with heavy metals, and clines in metal resistance can often be found over short distances that span contaminated and uncontaminated soils (Antonovics *et al.* 1971; MacNair 1981). An example involving *Agrostis tenuis* and *Anthoxanthum odoratum* from Hickey and McNeilly (1975) is given in Fig. 4.3, and illustrates the sharp change in resistance that can occur between a mine and a pasture area. Differences between resistant and susceptible populations usually have a genetic basis, although recent work on cadmium resistance (Baker *et al.* 1986) suggests that low levels of resistance may be inducible. Resistance is often considered to be specific to a particular metal (Turner 1969) although low levels of non-specific resistance have been described (e.g. Symeonidis *et al.* 1985), and some taxa may possess a generalized low level of resistance (Hutchinson 1984).

Localized genetic differences in plant populations have also been described for other environmental stresses. Silander and Antonovics (1979) examined three subpopulations (dune, swale, salt marsh) of the coastal grass, *Spartina patens*, along a 200 m transect. Experiments where subpopulations were grown from tillers under the same conditions (i.e. a 'common garden' experiment) provided evidence for genetic divergence in morphometric and physiological traits. The dune population was more resistant to drought, salt, and low nutrient conditions than the marsh population, and the swale population was intermediate. Competition studies indicated that selection in the dunes

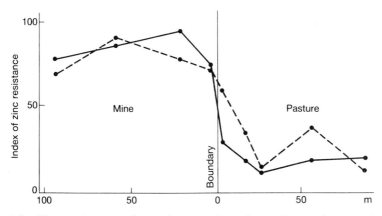

Fig. 4.3. Zinc resistance of populations of *Anthoxanthum odoratum* (solid line) and *Agrostis tenuis* (dashed line) at the mine boundary at Trelogan, Wales. The *A. tenuis* values have been multiplied by three. (After Hickey and McNeilly 1975.)

operated mainly via density-independent effects, while density-dependent effects were important in the swale and marsh populations. Antlfinger (1981) examined the shrub, *Borrichia frutescens*, which grows on a salinity gradient stretching from a high marsh environment to the edge of a salt flat where there are only a few stunted plants. Collecting the seeds of plants from high and low salt locations in the gradient, she showed differences in physiological traits between locations, even though there was substantial gene flow since locations were only 2 to 5 m apart. Ashenden *et al.* (1975) collected seed from cocksfoot (*Dactylis glomerata*) plants along a 130 m transect spanning different soil moisture levels across a dune and grew them under different water regimens. There was an inverse relationship between relative growth rate at the lowest water regimen and water retention at the sites, indicating localized physiological adaptation. Finally, populations of the annual Carolina cranesbill, *Geranium carolinianum*, near a power station that discharged SO_2 from a small smoke stack were significantly more resistant (Fig. 4.4) to SO_2 than those from rural areas, as measured by mean leaf necrosis response (Taylor and Murdy 1975). There was a great deal of variability within populations and the regression of the

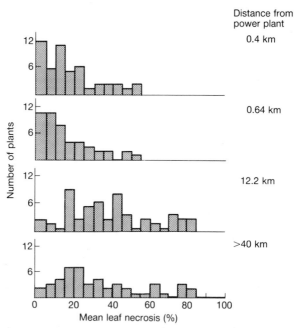

Fig. 4.4. Effects of sulphur dioxide (0.8 ppm ± 0.05) on pollution resistance as measured by leaf necrosis in populations of *Geranium carolinianum* collected different distances from a power plant. (After Taylor and Murdy 1975.)

offspring upon the mid-parent values gave a heritability around 50 per cent (Taylor 1978).

Examples of localized genetic differences in animal populations appear to be less common. Bradley (1978, 1981) described genetic variation for recovery from temperature shock in the copepod, *Eurytemora affinis*, from a region where populations are exposed to extreme temperature variation. Recovery times were correlated with survival at high temperatures. Heritability estimates for males were in the range 0.7 to 0.9 while the heritabilities for females ranged from 0.1 to 0.4. There was evidence of localized differences in high temperature resistance. Testing progeny from field samples collected near a power plant that discharged heat showed that males originating from sites affected by the discharge had higher heat resistance than males from other sites. These genetic differences contrasted with seasonal variation in temperature resistance that was mediated by physiological acclimation rather than genetic changes, even though the seasonal temperature change was greater than temperature variation at the power plant. It seems that copepods can increase temperature resistance by acclimation, but this process requires gradual temperature changes that do not occur during heat discharges. This was confirmed by a selection experiment where copepods were held in temperature regimens that cycled rapidly (1°C per day) or slowly (1°C per week): genetic changes were relatively more important under the rapidly changing regimen (Ketzner and Bradley 1982).

Localized genetic variation in ethanol resistance was described in a winery population of *D. melanogaster* by McKenzie and Parsons (1974*a*). The progeny of inseminated females collected inside and outside the winery cellar were found to differ for larval ethanol resistance, the more resistant isofemale strains coming from inside the cellar. This microdifferentiation was demonstrated over a 3 year period and has been found in subsequent collections (J. A. McKenzie, personal communication). Selective agents that might be associated with the differences in ethanol resistance remain unknown, because wine seepage resources inside the cellar do not contain unusually high levels of ethanol or other alcohols compared to other *D. melanogaster* resources (Gibson *et al.* 1981; McKechnie and Morgan 1982). Subsequent experiments with flies from the same population have revealed differences in the adult ethanol resistance of stocks originating from a grape residue pile and from traps with oranges and apples a few meters away, although differences were not found between stocks originating from the fruit traps and the cellar (Hoffmann and O'Donnell 1990). A genetic difference between the grape pile and fruit trap stocks was confirmed by the absence of maternal effects in reciprocal crosses. Finally, Gibson and

Wilks (1988) have found differences in the adult ethanol resistance of stocks from inside and outside another winery cellar where fortified wines were produced. Fortified wines have a high ethanol content, and Gibson and Wilks showed that the ethanol content of *Drosophila* food resources inside the cellar was higher than that of resources outside the cellar, suggesting that ethanol may have been the selective agent in this case.

As a final example of microgeographic variation in animals, Klerks and Levinton (1989) described spatial variation in the resistance to cadmium and nickel of the freshwater invertebrate, *Limnodrilus hoffmeisteri*. Stocks originating from polluted sites in a cove were more resistant to sediments containing these pollutants than stocks from a control site, 2 km away (Fig. 4.5). There was also spatial variation in resistance within the cove which matched variation in levels of pollutants. Artificial selection for resistance on individuals from the control site led to a rapid selection response, and heritability estimates in this population ranged from 0.59 to 1.08.

What conclusions can be reached? As for most quantitative traits, genetic variation for stress resistance can be readily found in populations, at least when they are tested under laboratory conditions or (in the case of crop plants) in agricultural settings. However, the extent to which genetic factors contribute to the phenotypic variance under natural conditions is largely unknown. Genetic differences in stress

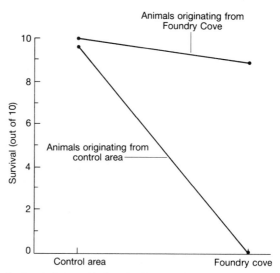

Fig. 4.5. Survival of *Limnodrilus hoffmeisteri* from Foundry Cove and a control area when exposed to sediments from each location. (Modified from Klerks and Levinton 1989.)

resistance commonly occur at the geographic level, and follow climatic variables, indicating that populations have undergone past episodes of selection for resistance. There are also examples where genetic differences in resistance can be found over short distances and between seasons. In agreement with Ehrlich and Raven (1969), these examples suggest that differentiation may develop between sites despite considerable gene flow, and suggest that strong selection for stress resistance occurs in natural populations. Both the variation in resistance in space and over time also suggest that there are 'trade-offs' such that increased stress resistance leads to decreased fitness under non-stressful conditions; this will be discussed further in Chapter 7.

4.5 Variation in stress evasion

Changes in an organism's life-history can minimize the time spent in an unfavourable environment and therefore allow evasion of environmental stress. An extreme example of such evasion is seen in organisms which enter dormant stages in response to stressful conditions. An organism may match its life-history to environmental conditions by completing development in favourable conditions and passing through unfavourable periods in a dormant stage. This can be a response to many conditions where energy acquisition becomes severely curtailed, such as a reduction in food supply or extreme climatic conditions that restrict growth and reproduction. This stress response is found in many species of plants and invertebrates that complete development in the favourable period directly following soaking rains in arid regions, and in organisms that complete development in the short growing season found at high latitudes. Dormant stages have very low metabolic requirements and allows conservation of energy. Even though these stages are stress resistant, dormancy is considered a form of stress evasion since dormant organisms do not grow or reproduce.

Environmental stress can also be evaded by switching resources, so that the stressful conditions are eased or cease to exist. Some animals switch to a new food supply to avoid starvation: many phytophagous insects expand the breadth of their diet as they become starved (Parker 1984), and scorpions feed less, eat smaller prey, and tend to feed more on prey species that are noxious or dangerous as food stress increases (Polis 1988). Plants may undergo morphological and physiological changes in order to utilize another resource. For example, plants that normally produce leaves adapted to a high light intensity may evade a low light stress by producing shade-adapted leaves.

A third way in which organisms may evade a stress is by moving away and finding a habitat where the stress is absent. Some mosquito strains are repelled by DDT and seek out alternative habitats where DDT is not found (Georghiou 1972). Many wingless insect species faced with deteriorating habitats are capable of developing wings and moving to more favourable habitats (Harrison 1980). Some plants can modify the growth of their roots when they encounter soil zones poor in nutrients, and project their roots into nutrient rich zones (Grime *et al.* 1986).

There is evidence for genetic variation for these stress evasion mechanisms in natural populations, and this is particularly true in the case of evasion by dormancy. The literature review by Tauber *et al.* (1986) on life cycle variation in insects indicates that geographic variation in diapause characters is usually under genetic control. Some of this variation has been related to environmental stresses experienced by populations. For example, Tauber and Tauber (1982) described a population of the lacewing, *Chrysopa*, in Alaska that was adapted to a very harsh physical environment with a short growing season. Diapause and reproduction were controlled by photoperiodic requirements, which ensured reproduction and development after the risk of severe cold spells had passed. Ovarian development proceeded rapidly to counter the short growing season. In contrast, termination of diapause and initiation of reproduction in another population from California, where the climate was more favourable, depended on the availability of prey rather than climatic factors.

Many genecological studies with plants also suggest genetic variation for stress evasion by dormancy at the geographic level. For example, the work of Clausen *et al.* (1948) indicated ecotypic variation in vegetative dormancy in *Potentilla glandulosa* and *Achillea millefolium*. High-altitude populations exposed to periods of climatic stress in winter were winter dormant, while some low-altitude populations were summer dormant. Other examples are described in Heslop-Harrison (1964).

Genetic variation of dormancy has been found within natural populations. Variation in insect diapause traits, such as responses to factors inducing and terminating diapause or influencing the duration of diapause, is often continuously distributed. These traits may be changed readily under artificial selection. For example, in the milkweed bug, *Oncopeltus fasciatus*, artificial selection experiments and parent–offspring regression indicated that a high proportion of the variance for the critical period inducing diapause was genetic, with heritability estimates around 70 per cent, and this enabled selection of non-diapausing strains under a particular light regimen in a few generations (Dingle 1978). Similarly, selection for diapause in a population of the pitcher plant mosquito, *Wyeomyia smithii*, changed the frequency of non-diapausers from about 50 per cent to less than 10 per cent in a few generations (Istock *et al.* 1976). Selection for shortened diapause in the lacewing, *Chrysopa carnea*, reduced the time between emergence and oviposition from 69 days to 7 days in only four generations (Tauber and Tauber 1982).

Variation in diapause characters within insect populations often shows discontinuities. There are many examples (Waldbauer 1978), although the genetic basis of the discontinuous variation has only been examined in a few cases. In the cabbage root fly, *Delia radicum*, populations can be grouped into early-, intermediate-, and late-emerging types, the early type emerging from diapause within 14 days and the later type after more than 100 days at 20°C (Finch and Collier 1983). Intermediate types had a proportion emerging within 14 days and a proportion of later emergents. This variation was genetically based because differences between populations persisted under laboratory culture, and because early-, intermediate-, and late-emerging flies could be selected from a parental population in one generation (Fig. 4.6). Discontinuous variation in the length of diapause may span a period of more than one year, in which case a small percentage of the population remains in diapause while most of the population becomes reproductively active. For example, most individuals of the sawfly, *Perga affinis*, feeding on eucalyptus leaves were univoltine, but a few remained in diapause in the prepupal stage for 12, 24, or 36 months (Carne 1962). This type of variation is common in insects, although its genetic basis has been little studied because of the long-term nature of breeding experiments.

While the insect work has indicated the importance of genetic factors,

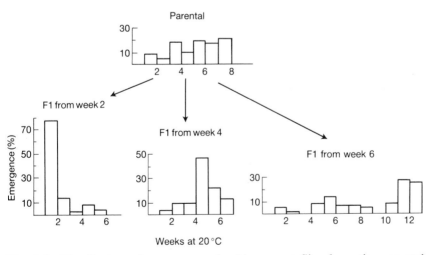

Fig. 4.6. Distribution of emergence of cabbage root flies from the parental generation and from the F1s reared by selecting from the parents emerging after 2, 4 or 6 weeks. (Modified from Finch and Collier 1983.)

variation in dormancy within plant populations may be largely environmentally determined at least in some cases. For example, Arthur *et al.*
(1973) examined variation in germination time in the poppy, *Papaver*
dubium. Seeds germinate in either spring or autumn. Progeny from
groups of plants from these two seasons did not show any tendency to
germinate in the same season as their parents, indicating that variation in
germination time was largely environmental.

While dormancy is a drastic form of stress evasion, less dramatic
changes in life-history traits may also contribute to evasion. Many
aspects of life-history theory have been concerned with predicting the
types of changes that may occur in response to environmental stress at
different life cycle stages (reviewed in Stearns 1976; Partridge and
Harvey 1988). For example, Charnov and Schaffer's (1973) model
comparing annual and perennial populations incorporated differences
in juvenile and adult survival to predict that organisms should breed
repeatedly when adult survival is high relative to juvenile survival. This
requires different stress levels at different life cycle stages. By remaining
in the adult stage for longer, individuals evade the greater stress levels
experienced at the juvenile stage.

Environmental stress is also a component of models concerned with
responses to environmental fluctuations. These models predict that
organisms should maximize their geometric mean fitness rather than
their arithmetic mean fitness (Gillespie 1977), implying that periods of
stress will make a disproportionately large contribution to an organism's
overall phenotype if extremes are selected during rare stressful conditions. There is some evidence that phenotypes in populations reflect
the effects of rare periods of stress. For example, mean clutch size in
great tits (*Parus major*) is smaller than the number of young that can be
raised by adults in average years (Boyce and Perrins 1987). Individuals
producing large clutches were selected against in poor years for survival
of young; a small clutch size helps birds to evade stressful conditions.

Finally, some aspects of life-history theory have attempted to predict
phenotypes that are likely to evolve in different types of habitats (Section
2.4). Some of the phenotypic classifications are tied to environmental
stress evasion, particularly those associated with the concepts of *r* and *K*
selection. Populations that are *r* selected occupy areas where physical
factors are limiting and rapid rates of increase are favoured, in order to
take advantage of the limited resources available (i.e. stress evasion). On
the other hand, *K* selection occurs where biological interactions predominate and stress evasion is not selected. Phenotypes predicted by *r*
selection (early reproduction, large clutches and reproductive effort,
small offspring, low assimilation efficiency, short generation time)

enable organisms to complete growth and reproduction in short favourable conditions and evade stressful periods.

Many experiments have demonstrated genetic variation in life-history
traits (Mousseau and Roff 1987), but few studies have considered this
variation in relation to environmental stress. Newman (1988*a,b*)
described an example in the toad, *Scaphiopus couchii*, which breeds in
ponds of uncertain duration. Larvae that developed slowly died when
ponds dried too quickly even though such larvae were larger at metamorphosis. Stress could therefore be evaded by a rapid development
time. Newman found high levels of additive genetic variance in development time within a population of *S. couchii*, and suggested that genetic
variation may persist because spatial selection favours different genotypes in different ponds. Variation in development time was also
influenced by the environment, because toads from all sibships
developed faster in ponds of short duration. The fact that faster
developers tended to have smaller body sizes suggests that stress evasion
may entail a cost, since larger toads will have a greater reproductive
output. This will be discussed in Section 7.6.

The abnormal abdomen (*aa*) polymorphism in *Drosophila mercatorum* (Templeton and Johnston 1982, 1988) provides another
example where life-history variation has been related to stress. Flies with
an abnormal abdomen had a prolonged larval development stage compared to non-*aa* flies, but their reproductive maturity was speeded up by
2 to 3 days. These early maturing females had increased ovarian output
but decreased longevity in the laboratory. Expression of these phenotypic effects is associated with insertions in the rDNA coding for ribosomal genes on the X chromosome. Phenotypic expression also requires
the absence of a compensatory mechanism associated with X-linked
genes. In Hawaii, *D. mercatorum* adults normally die within a few days
under dry conditions in the field because of desiccation stress, and live
much longer under humid conditions. These environmental effects on
longevity mean that dry conditions favour *aa* flies which mature earlier
and have a fecundity advantage, while humid conditions favour non-*aa*
flies, with their increased longevity. These fitness differences help to
explain spatial and temporal variation in the frequency of *aa* X chromosomes along an altitudinal gradient (Fig. 4.7). Dry conditions normally
occur at low altitudes and humid conditions are normal at high altitudes;
this is consistent with the spatial pattern of *aa* in 1980, when *aa* was
relatively more common in the low altitude site (IV). During a dry year
(1981), the frequency of *aa* increased at all sites except IV, while the
reverse occurred in a wet year (1982). The weather was normal again in
1984 and 1985 except for high humidity at site IV, and this was reflected

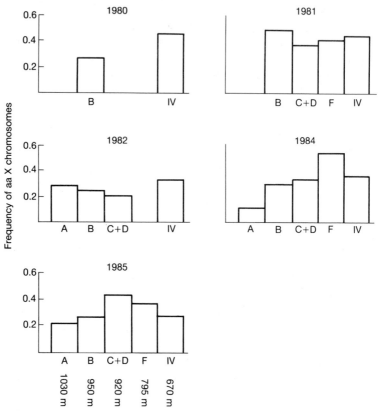

Fig. 4.7. Frequencies of X chromosomes allowing expression of abnormal abdomen in *Drosophila mercatorum* from five locations at different altitudes sampled over 5 years. (Based on data in Templeton and Johnston 1988.)

in *aa* frequency changes. Genetic variation influencing life-history could therefore be related to changing climatic conditions.

There is little information on genetic variation for stress evasion by resource switching. Darwin's finches (*Geospiza fortis*) on Daphne Major Island in the Galapagos provide an example where a starvation stress was evaded via a morphological character (body size). A drought in 1977 was associated with 85 per cent mortality in the population. Survivors showed an increase in beak and body size (Boag and Grant 1981), and this pattern of selection was repeated in two subsequent periods of climatic stress (Price *et al.* 1984). Large birds were favoured because they were able to crack the large and hard seeds that formed the main food source available during periods of drought; they were therefore able to evade starvation stress by access to a different food supply.

Genetic changes were involved because morphological characters in Darwin's finches have a high heritability, including those that changed as a consequence of selection (Boag and Grant 1978).

Genetic variation for stress evasion by moving away has been described in insect populations that are polymorphic for flight ability. Examples include several species of water striders of the genus *Gerris* (Vepsalainen 1978). Populations consist of long- and short-winged individuals. These variants are produced by different genotypes in some species since progeny have the same morph as the parents. Variants are, however, environmentally determined in other species, and the environment may also influence the expression of genetic variation. In *G. lacustris*, for example, some long-winged individuals produce short-winged offspring under summer conditions, but not at other times. Short-winged individuals are unable to fly and do not produce flight muscles. The distribution of wing morphs indicates that long-winged individuals tend to occur in areas that are unfavourable, such as ponds that dry up frequently and require repeated colonization, while short-winged individuals tend to be favoured in permanent habitats. Polymorphisms for flight ability are found in many other insects, including aphids, beetles, planthoppers and crickets, although a genetic basis for such polymorphisms has only been established in a few cases and they may be largely environmentally determined in aphids (Harrison 1980).

The above discussion is not meant to provide an exhaustive review of genetic variation in stress evasion, but gives an indication of traits that may be involved in evasion, and the occurrence of genetic variation for these traits within and between populations. The presence of substantial genetic variation for many stress evasion traits within populations raises the question of how such variation persists in nature (Section 5.7). Stress evasion involves a general cost, forfeiting resources when organisms become dormant or move away, instead of directly counteracting the effects of a stress. The patchy occurrence of stressful conditions may provide situations that sporadically favour phenotypes associated with stress evasion.

4.6 Genetic variation for phenotypic plasticity

Phenotypic plasticity was defined by Bradshaw (1965) as:

the amount by which the expressions of individual characters of a genotype are changed by different environments.

Plasticity covers a number of processes (Section 1.3), including reversible changes in physiological processes in response to environmental conditions, such as changes in metabolic rate in animals or the closure of stomata in plants, as well as more complex processes, such as acclimation to environmental conditions where physiological stress resistance is increased after exposure to periods of sub-lethal stress. Finally, plasticity encompasses phenotypic changes during development that are triggered by the environment but are often irreversible (developmental conversion), such as the production of dormant stages in the life cycle of insects.

Phenotypic plasticity is extremely common in plants, and is often expressed in stressful environments. Some of the more spectacular examples in Bradshaw's (1965) review are the production of cleistogamous flowers (which do not open and are self-fertilized) in dry periods compared to open flowers in other periods, and the loss of leaves in response to climatic stresses. The degree of expression of phenotypic plasticity is under genetic control; populations and species show different levels of plasticity for the same characters in response to the same environmental variables (Bradshaw 1965; Schlichtling 1986). Genetic assimilation experiments (discussed in Section 5.2) also indicate that phenotypic plasticity can be altered by selection: phenotypes that are initially expressed only after exposure to a stress can be selected so that they are always expressed (Waddington 1957), implying a change in plasticity levels.

Animals tend to have lower levels of morphological plasticity than plants, but many of the physiological processes which may be influencing stress resistance show a high degree of plasticity. Ectotherms often undergo changes in metabolic rate in response to temperature, and this can be modified by acclimation. The respiration rate of individuals acclimated at warm temperatures slows down at intermediate temperatures, while the reverse occurs when they are acclimated at cold temperatures (Bullock 1955; Schmidt-Nielsen 1984). For example, high temperature increases metabolic rate in *D. melanogaster*, and the metabolic rate of females grown at 25°C and 30°C was lower than that of flies grown at 15°C when tested in the 20 to 30°C range (Hunter 1964). This ensures that organisms maintain a similar metabolic rate over a range of temperatures, avoiding energy wastage at high temperatures, and ensuring sufficient metabolism at low temperatures for normal functioning. The ability of animals to undergo acclimation is under genetic control; there are differences in the acclimatory abilities of related species. For example, metabolic acclimation similar to that found in *D. melanogaster* occurred in two other *Drosophila* species (*D. immigrans, D. hydei*) but was not detected in *D. willistoni* when they were cultured under similar conditions (Hunter 1966, 1968).

Natural selection may increase or decrease the level of plasticity in a trait. For traits closely related to fitness, such as viability or fertility, selection will tend to decrease plasticity as long as this decrease is not associated with low levels of viability or fertility. This is often referred to as selection for increased fitness homoeostasis, because selection favours genotypes that express high levels of fitness under a range of environmental conditions. Fitness homoeostasis may, in turn, be associated with increases in the plasticity of underlying traits that counter the effects of environmental changes. For example, some plants can maintain growth and reproduction under variable light conditions by producing different leaves adapted to shady or exposed conditions. This represents an example where increased plasticity in a morphological trait leads to fitness homoeostasis. Similarly, an insect that alters its cuticle composition to prevent desiccation under conditions of low humidity shows a high level of plasticity in cuticle development that helps to ensure continued survival and reproduction.

If species differences in plastic responses are adaptive, they should be relatable to habitat differences. Organisms living in variable environments should exhibit an increased ability to counter the effects of environmental changes, and thereby minimize variation in metabolic rate and other physiological processes. Differences in the acclimatory response of *Drosophila* species to temperature (Hunter 1964, 1966, 1968) fit this expectation. *D. melanogaster* is widely distributed and occurs mainly in the temperate zone, where temperature fluctuations are large. Temperature acclimation is therefore expected in this species, as well as in *D. immigrans* and *D. hydei*, which are also cosmopolitan species (Parsons and Stanley 1981). In contrast, a species not showing acclimation for metabolic rate (*D. willistoni*) is more restricted in its distribution and will encounter smaller temperature fluctuations in its tropical environment.

Environmental variability has also been associated with phenotypic plasticity at the interspecific level in plants. Variability in the physical environment of early successional species tends to be greater than that of late successional species, and this should be reflected in greater flexibility of physiological processes in the former group. Bazzaz and Carlson (1982) found that plants from early successional habitats showed greater photosynthetic flexibility when grown in light and shade environments than late successional species: early succession annuals could lower their photosynthetic rate to the rates characteristic of shade-adapted plants. This increased physiological plasticity is likely to be an evolved response to variability in light conditions.

Plasticity levels may also reflect the levels of environmental stress experienced by species in their natural habitats. Comparisons of plants from stressed and favourable habitats grown under the same conditions show that species from favourable, productive habitats tend to have high levels of morphological plasticity when exposed to an environment with patchy conditions, whereas species from unproductive habitats tend to have low levels of morphological plasticity (Grime *et al.* 1986). This can be illustrated by the response of six bryophyte species to patches of unfiltered light (Fig. 4.8). Stress resistance in bryophytes, as in many other plants, is negatively correlated with growth rate under favourable conditions (Chapter 6), and bryophytes from harsh habitats (shaded limestone outcrops) had slower growth rates than species from more favourable, mesic habitats. Growth rate was correlated with the degree of responsiveness to light patches, in that slow growing species concentrated less of their new growth in these patches, reflecting a decreased level of morphological plasticity.

Plasticity in physiological traits has been related to interspecific variation in stress resistance in intertidal animals (Branch *et al.* 1988). Intertidal species from areas normally low in food have low oxygen

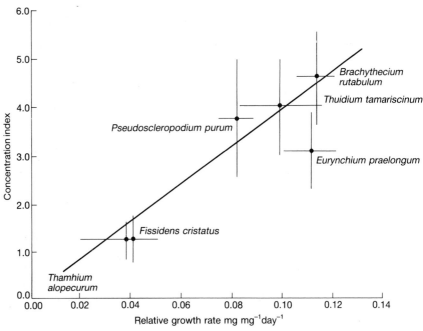

Fig. 4.8. Comparison of the ability of six bryophytes grown in a constant mosaic of high and low irradiance to concentrate their biomass in areas of high irradiance. Concentration indices are plotted against relative growth rate under uniformly high irradiance. 95% confidence limits are represented by the horizontal and vertical lines, and the fitted line from a linear regression is also given. (After Grime *et al.* 1986.)

consumption and do not greatly modify their metabolic rate when they are provided with a variable food supply. In contrast, species from areas where food is abundant have higher oxygen consumption but are able to lower this as food becomes scarce, indicating an increased level of plasticity for this physiological variable.

These findings indicate that differences in the levels of plasticity displayed by species can be associated with the types of environments that they experience. Such an association has also been found in a few studies at the intraspecific level. Billings *et al.* (1971) compared the effects of temperature acclimation on metabolic rate in alpine and arctic ecotypes of the tundra plant, *Oxyria digyna*. Maximum net photo-synthetic rates were lowered to a greater extent in arctic populations than in alpine populations after warm acclimation, indicating an increased ability of alpine populations to compensate for temperature changes. Both ecotypes showed similar acclimatory responses for dark respiration. Daily temperature fluctuations are greater in the alpine environment than in the arctic environment, and select for the ability of these populations to change their photosynthetic rates to compensate for changes in temperature.

Environmental stress and morphological plasticity have also been associated at the intraspecific level. McGraw (1987) used reciprocal transplants to examine variation in two ecotypes of the dwarf shrub, *Dryas octopetala*, which can be distinguished on the basis of mor-phology. One ecotype grows on unfavourable fellfields that are low in nutrient availability, exposed to high winds, and have a sparse vegetation cover. The other ecotype grows in snowbeds which have high nutrient availability and support a more dense vegetation, but have a short growing season and are covered with snow in winter. Comparison of two sites representing these habitats that were less than 100 m apart revealed that ecotypes did not show mortality on the sites where they originated. The fellfield type showed little initial mortality on the snowbed site, but survival gradually declined over a 7-year period. In contrast, plants originating from the snowbed suffered heavy initial mortality on the fellfield, but this was reduced in later years due to a plastic response, and these plants closely resembled the fellfield ecotype after one year. These ecotypes showed different levels of phenotypic plasticity, as well as being relatively better adapted to the conditions from which they originated. The ecotype from the less favourable fellfield site had a relatively lower level of morphological plasticity.

Differences in plasticity levels between species and populations are usually explained in terms of natural selection adjusting plasticity levels in response to the degree of environmental variability experienced by populations. However, increased stress resistance may also reduce the level of plasticity if the same mechanism controls the plastic response and genetic variation in stress resistance, and this could account for the association between resistance and plasticity discussed above. There is evidence at the enzyme level that similar changes in enzyme concentration and isozymes account for species differences and acclimation responses (Hochachka and Somero 1984) that could form the basis of such an association. In addition, kinetic parameters of enzymes from clones of plant species from different environments often show different levels of plasticity. For example, while Simon *et al.* (1986) found that the V_{max}/K_m ratios for malate dehydrogenase in *Viola* were higher at assay temperatures closely related to normal growth conditions of the clones (Section 3.1), the response of clones to assay temperature also differed. Clones from colder climates showed a more rapid increase in K_m with increasing temperature than clones from warmer climates. Direct evidence that stress resistance and acclimation responses are associated comes from *D. melanogaster* lines selected for increased desiccation resistance. A non-lethal period of desiccation increases subsequent desiccation resistance in *D. melanogaster* (Hoffmann unpublished). When the lines selected by Hoffmann and Parsons (1989*a*) were acclimated by exposure to such non-lethal stress, the desiccation resistance of control lines increased markedly, but there was little change in the resistance of the selected lines (Table 4.3).

We have so far concentrated on plasticity in stress resistance traits, but traits involved in stress evasion also show phenotypic plasticity. Much of the discussion in this area has focused on plasticity in the size of propagules produced by organisms from stressful and favourable environments (Capinera 1979; Crump 1981; Kaplan and Cooper 1984). Many insects, plants and amphibians occupying unpredictable habitats produce propagules of variable size rather than a single optimum size, and the range of variation produced by species is under genetic control. Comparison of species indicates that those occupying environments with intermittent stress periods produce more variable propagules. For example, in tropical tree frogs of the genus *Hyla*, clutches from species breeding in temporary ponds tended towards platykurtosis (fewer items near the mean), while those breeding in permanent ponds tended towards leptokurtosis (more items near the mean), suggesting that the latter were selected more strongly for a single optimum size while the former were selected for greater variability (Crump 1981). The production of variable progeny may ensure that at least some are the appropriate size for prevailing conditions.

Plasticity in stress evasion traits may arise as a consequence of responses to environmental stress. For example, Stearns (1983) induced stressful conditions in mosquitofish by varying temperature and the

Table 4.3. Desiccation resistance of selected and unselected lines of *Drosophila melanogaster* with and without a prior acclimation period. Lines were selected for increased desiccation resistance in the absence of acclimation as described in Hoffmann and Parsons (1989*a*). Flies were acclimated by placing them at <5% humidity for nine hours before allowing them to recover for nine hours.

	Desiccation resistance*	
	Non-acclimated	Acclimated
Selected lines		
Line 1	23.6 (1.7)	24.8 (1.7)
Line 2	22.1 (1.1)	23.1 (1.5)
Line 3	24.2 (1.3)	24.9 (2.0)
Control lines		
Line 1	16.2 (1.1)	19.9 (0.9)
Line 2	17.7 (1.1)	21.4 (2.1)
Line 3	15.0 (1.4)	19.2 (2.1)

*Mean time taken for 50% of the females to die (in hours), based on six replicate groups of 20 females. Numbers in brackets indicate standard deviations.

degree of crowding. He found that age at maturity and, to a lesser extent, body length at maturity changed with environmental conditions. Stearns proposed that plastic changes should depend on the mortality levels experienced by organisms. If they experience high pre-maturation mortality because of stressful conditions, they should reduce their age at maturity at the expense of a reduced length at maturity, while the reverse should occur if they experience low mortality. The ability of species to exhibit these plastic changes has presumably been selected, and will therefore depend on the conditions that they normally experience. Supporting evidence comes from Kaitala's (1987) investigation of reproduction and longevity in the water strider, *Gerris thoracicus*, under conditions of variable food availability. Females had a high reproductive output and a short lifespan when food was abundant, and showed decreased reproductive output but increased longevity when food was scarce. This species lives in small rock pools that often dry out and provide a variable food supply; life-history changes should therefore help to increase survival when food is scarce and maximize reproduction in the absence of stressful conditions. In contrast, another species of

water strider (*G. lacustris*), which lives in more permanent pools
showed decreases in reproduction and longevity when food was scarce.
The life-history of *G. thoracicus* was therefore more plastic, presumably
because natural selection has favoured genotypes that allow life-
histories to be altered in response to stressful conditions.

In summary, the empirical data indicate that phenotypic plasticity for
stress response traits is under genetic control, and there are several cases
where plasticity levels appear to have been modified by natural selection.
Organisms that rarely experience stressful conditions seem to show a
reduced ability to acclimate compared to organisms from variable
environments, and levels of morphological plasticity may be lower in
plants from stressful environments than in plants from favourable
environments. More experiments are required that extend these interspecific patterns to the intraspecific level. In particular, the effects of
fluctuating environments or continuously stressful environments on
genetic variation for plasticity levels needs to be examined.

4.7 Conditions favouring phenotypic plasticity

We have seen that seasonal, local and geographical variation in stress
response traits may involve genetic variation and/or plasticity for stress
responses. This raises two questions about conditions favouring plastic
responses to environmental stress:

(1) What factors influence the relative contribution that plastic and
    genetic factors make to phenotypic variance in stress responses
    within populations?
(2) What factors favour population responses to environmental stress
    by genetic changes or plastic changes?

It is not clear whether genes influencing the degree of plasticity of a trait
and genes influencing resistance or evasion should be treated as independent of each other. The evidence discussed above suggests that high
levels of plasticity may be associated with low levels of stress resistance,
so that more of the phenotypic variance in populations with low levels of
stress resistance could be associated with plasticity. There is also some
evidence that species with high levels of genetic variation have low levels
of plasticity. For example, Jain (1979) compared three congenic species
pairs for plasticity and genetic variation, and found that, within each
pair, the species with the highest level of genetic variation as measured
by protein electrophoresis had relatively lower levels of morphological
plasticity in common garden experiments with variable soil type,
moisture and photoperiod regimens.

Naïvely, we might expect that phenotypic variation in stress responses within populations, as well as population changes under stressful conditions, would largely involve plastic changes, because selection would favour a single genotype able to undergo an infinite number of compensatory plastic changes to ensure that fitness remains high under all environmental conditions. However, this is clearly not the case—there are plasticity differences within and between species which can often be related to environmental variation, as discussed above, suggesting that high levels of plasticity are favoured under some conditions and selected against under others. In addition, levels of genetic variation for stress response traits are high in many populations (Section 4.3). Several authors have suggested factors likely to influence the contribution of plasticity and genetic variation to phenotypic variation within and between populations, and these are briefly discussed with respect to stress response traits. We assume here that high levels of plasticity are associated with fitness homoeostasis because these reflect compensatory physiological and morphological changes that counter the effects of environmental stress.

4.7.1 Frequency of stress

Bradshaw (1965), Levins (1968), and Lynch and Gabriel (1987), among others, have suggested that the importance of plasticity will depend on the frequency of recurrent environmental changes. An organism's stress response will involve phenotypic plasticity if the duration of recurrent environmental changes is less than the generation time. Reversible plastic responses will be favoured if the environment changes often, while irreversible plastic changes early in development may be favoured if the initial environmental conditions encountered by an organism persist for some time. Stress periods during a single generation could, theoretically, select for genotypes with increased stress resistance. However any genetic changes in a population may be reversed if environmental conditions do not impose a stress during the next generation, because genes favoured under stressful conditions are not likely to be favoured under favourable conditions (see Chapter 7). In contrast, genes increasing plasticity levels would always be favoured, so that high levels of plasticity should be selected when environmental changes are frequent.

Bradshaw (1965) has argued that rapid environmental changes will be commonly experienced by plants, which cannot move away from stressful conditions, and he suggested that periods of recurrent selection in plants will often occur within one generation. In addition, seeds from plants growing in one environment will often be exposed to a different environment, because of the sessile nature of plants and the high levels of spatial and temporal heterogeneity in most terrestrial environments.

Bradshaw's verbal argument has been criticized by Orzack (1985) who showed that genotypes with increased fitness homoeostasis in different environments may or may not be favoured in environments that change at intervals of less than the generation time. Orzack considered a model of environmental variability in an age-structured population where the relative fitness of homozygotes switches between environments, and where heterozygotes have intermediate fitness. He found that heterozygotes are not always favoured in this model, even though they have reduced fitness homoeostasis when averaged over environments. Orzack's criticism may be relevant to the specific case where heterozygotes have higher fitness homoeostasis and when the fitness of heterozygotes is always less than that of one of the homozygotes. However, the reduced fitness of the heterozygotes implies that there is a cost to increased plasticity (see below). Bradshaw (1965) was not considering this specific case and his arguments are applicable when there is no cost or when increased plasticity is not associated with heterozygosity. Furthermore, the importance of within-generation variation in favouring plastic genotypes is supported by Lynch and Gabriel's (1987) quantitative analysis of genotypes with different tolerance curves, where genotypes with broad tolerance curves (high levels of plasticity) were favoured when environments varied in time.

4.7.2 Microspatial environmental heterogeneity in stress levels

Stress responses can only involve phenotypic plasticity if spatial environmental variation is encountered by a single individual (Sultan 1987). For example, roots may encounter different soil conditions depending on depth, and leaves may encounter a variety of shade conditions depending on their position in a canopy. This type of variation occurs within most plants and means that individuals encounter a range of stress conditions requiring plastic responses. Animals can avoid such spatial patchiness by moving, and this factor may contribute to the lower morphological plasticity exhibited by animals.

4.7.3 Magnitude of stress and physiological/genetic limits

Responses to an environmental stress may involve genetic changes when plastic changes are inadequate to counter the effects of a stress. For example, the evolution of insecticide resistance and heavy metal resistance in plants requires genetic changes because plastic changes are inadequate to counter the toxic effects of the chemicals. Insecticide resistance involving structural changes in enzymes targeted by the insecticide may be encoded at the DNA level so that there is only a limited possibility for plastic changes in resistance. In general, this factor will be important when genetic changes and plastic changes in stress resistance are based on different mechanisms.

Levins (1968) discussed the effects of stress on plasticity levels within a population by comparing the tolerance range of an individual organism to the environmental range that it experienced. He showed that a single phenotype would be favoured in a population occupying a variable environment when the environmental range encountered by the population is smaller than the tolerance range of an individual, or in other words when plastic responses of individual organisms can cope with the stress periods they experience. However genetic variation could persist if the population's environmental range exceeds the tolerance range of individual organisms because the plastic response becomes inadequate. This argument assumes the absence of genetic variation for plasticity that could lead to the evolution of expanded tolerance ranges.

Plasticity levels for stress evasion as well as for stress resistance may be insufficient to counter the effects of an environmental stress. For example, Newman (1988b) found plasticity for development time in the toad *Scaphiopus couchii*, which occupies temporary ponds (Section 4.5). Toads from ponds of short duration developed faster than those from more permanent ponds. There was no significant genetic variation for plasticity, suggesting that further responses to pond duration could not occur via increased plasticity. However there was evidence for genetic variation that could increase or decrease development times beyond the limitations imposed by plastic responses. Additive genetic variation for plasticity for life-history traits was also absent in milkweed bugs reared at different temperatures (Groeters and Dingle 1988).

4.7.4 Speed of change in environmental conditions

Plastic responses may be inadequate when an environmental stress arises rapidly. Organisms may have insufficient time to change their growth form or undergo physiological alterations to enable them to counter the effects of a stress. In particular, acclimation responses often develop slowly and require a long period of sub-lethal stress levels. This seems to explain Bradley's (1981) results for the copepod, *Eurytemora affinis*, discussed in Section 4.4. Bradley found that seasonal changes in temperature resistance were not genetically based, while microspatial variation in responses to heat discharges from a power station was genetically based, even though smaller temperature shifts were involved in the latter, presumably because the temperature changes at the power station were too rapid for acclimation to have much influence on stress resistance.

Plastic responses may also be inadequate when organisms under stress can take advantage of brief favourable conditions. Plants that have low levels of plasticity and maintain a large root mass under all fertility conditions may be at an advantage in soils of low fertility when nutrients occur in short pulses because the roots of these plants can take advantage of the pulses (Chapin 1980; Crick and Grime 1987). In contrast plants that can reduce their root mass when roots encounter low nutrient conditions and increase it in response to high nutrient conditions may not be able to develop their root system fast enough to take advantage of short pulses of nutrient. Crick and Grime (1987) provide empirical evidence supporting this possibility.

4.7.5 Reliability of environmental cues

The development of dormant life-cycle stages or resistant morphs involving relatively irreversible plastic changes often depends on stimuli such as photoperiodic cues that precede the environmental change. These stimuli need to be reliable if such plastic changes are to be favoured by selection. Lively (1986) considered a model where development can produce one of two variants with different fitnesses in a harsh and a benign environment. One variant always developed the resistant phenotype, another the susceptible phenotype, and a third plastic variant produced resistant or susceptible phenotypes depending on the environment. The plastic variant persisted in a population when it made the right choice more than 50 per cent of the time and when there was a cost to producing the resistant variant. Irreversible forms of plasticity may therefore be favoured when there are reliable cues for at least one of the environments, while genetic variation may persist in a population when cues are unreliable.

Lively (1986) discussed his model with respect to a population of acorn barnacles that can produce a hooded shell in response to attack by a predator. This morph is more resistant to attack but has lower fecundity and grows slower than morphs without hoods. Attack may be a poor predictor of future predation because of seasonally variable foraging by the predator. Gross (1984) considered light intensity as an unreliable cue that may select for low levels of plasticity for the development of leaves adapted to different light conditions. Plant species producing leaves adapted to low light conditions tend to show low levels of morphological plasticity compared to plants adapted to high light conditions. Gross argued that leaves adapted to high light intensities become maladapted in understorey plants as a canopy develops, even though they may have been beneficial early in the growing season. If leaves are maintained throughout the growing season then selection will favour low plasticity in shade plants because of the extended period of low light stress experienced by these plants. The environmental cues available during early development of the leaves are unreliable, so that genes decreasing plasticity will be favoured.

4.7.6 Cost of being plastic

Maintaining a plastic phenotype may entail a cost which leads to selection against genotypes with high levels of plasticity when variable environmental conditions do not occur (Heslop-Harrison 1964; Bradshaw 1965). One type of cost involves the genetic machinery required for the expression of a range of phenotypes (Williams 1966). This is probably small at the DNA level, but it may increase once the DNA is transcribed and translated. Under irreversible plastic changes early in development, only one set of genes is expressed so that costs should be minimal. When plastic changes are reversible, they may be associated with the production of additional gene products, although responses to environmental change can occur via the short term production of heat-shock proteins and other gene products. Costs may involve factors other than the energy required for producing additional mechanisms for coping with environmental change. These are discussed more extensively in Chapter 7 when we consider the performance of genotypes under stressful and optimal conditions.

The fact that many organisms are specialists rather than generalists and live in a restricted habitat rather than occupying a range of habitats suggests that costs are commonly associated with plasticity. The presence of genetic differences for plasticity levels within species also suggests selection against plastic genotypes. However, there appear to be some situations, such as in *Pinus monticola* (Rehfeldt 1979), where organisms occupying diverse habitats do so via environmental modification despite the presence of genetic variation within populations.

4.7.7 Historical constraints

Plastic responses to environmental changes may be complicated because they require mechanisms for detecting such changes as well as mechanisms for subsequently responding to them. The evolution of detection mechanisms may not occur readily, since they require a number of mutational changes, some of which may only be favoured when there is persistent genetic variation for stress responses within populations. This can be illustrated by a hypothetical example. Consider the case of variation in body colour that influences resistance to heat stress. Assume that an organism can have one of two morphs (light or dark) representing irreversible changes that may be controlled by genetic polymorphism or a plastic response involving developmental conversion, as in the Lively (1986) model discussed above. In a population initially fixed for a dark morph, will a light morph evolve by plastic changes when the population starts to experience periods of high temperature stress? Developmental conversion requires mutational change to produce the light morph, as well as mutations to produce a mechanism for the detection of temperature and an association between this mechanism and expression of the light morph. The evolution of developmental conversion in response to stressful conditions therefore requires mutational changes in several steps. Moreover, mutational changes producing the detection and conversion mechanisms will only be selected if they follow mutations producing the light morph.

Because several mutational changes are required in a sequence, irreversible plastic changes will only evolve when conditions that favour genetic polymorphism for colour persist for a long period of time. Plastic responses would not evolve if a long-term change in temperature conditions consistently favoured the more resistant morph, which would rapidly become fixed in a population, or if periods of stress were rare so that the light morph would only occur at a low frequency. Persistent genetic polymorphism within a population may only occur under restrictive conditions. For example, the Lively (1986) model predicts that genetic polymorphism will persist only with the long-term presence of patchy environmental conditions where both patch types occur relatively frequently.

4.7.8 Concluding remarks

This section can be summarized by Fig. 4.9, which lists factors that influence the importance of plasticity and genetic variation in phenotypic variation for stress response traits. Figure 4.9a considers whether the response of a population to environmental stress (or population differences associated with environmental variation) is likely to involve plasticity or genetic changes in stress responses. Common stress periods, slow environmental changes and resistance costs may favour plasticity, while large environmental changes and historical constraints may favour genetic changes. Responses to environmental change by phenotypic plasticity will also depend on the level of genetic variation and level of plasticity in the base population. The latter may be influenced by the factors listed in Figure 4.9b, which include frequency and magnitude of stress periods experienced by a population, microspatial environmental variation, costs of plasticity, and the reliability of cues for irreversible changes.

4.8 Summary

Genetic variation for body colour in ectotherms and endotherms has been associated with extremes of temperature. Climatic selection may

(a)

Response to environmental ──────────────────────── Response to environmental
change by phenotypic   change by genetic
plasticity   variation

high.. plasticity levels in base population .... low

low..................................... genetic variation for resistance/evasion .... high

low .... costs of phenotypic plasticity ....high

high .... costs of resistance/evasion ....low

common .... frequency of stress periods ....rare

slow .... speed of environmental change .... fast

small................................... magnitude of environmental change .... large

absent .... historical constraints ....present

(b)

High level of ──────────────────────────────── Low level of
plasticity   plasticity

low .... environmental predictability ....high
                                                (propagule production)

high.................................... reliability of environmental cues .... low
                                                (irreversible plasticity)

small .... magnitude of environmental variation .... large

common..................................... frequency of stress periods ....rare

present .... microspatial variation within individuals ....absent

low .... costs of phenotypic plasticity ....high

weak .... selection for canalization .... strong

Fig. 4.9. Factors that have been hypothesized as influencing phenotypic plasticity for stress response traits: (a) factors contributing to the importance of phenotypic plasticity or genetic variation in response to environmental change; (b) factors affecting the levels of phenotypic plasticity for stress response traits in a population.

act directly on this variation, although the temperature stress experienced by different morphs can be modified by behavioural responses.

In *Drosophila*, many studies have demonstrated genetic variation for resistance to a range of environmental stresses, but there is no information on levels of genetic variation under field conditions. Stress resistance variation has also been demonstrated in many agricultural plants and animals, but genetic information from natural populations is scarce. Several plant and animal examples indicate genetic differences in stress resistance over short distances despite gene flow. A few examples also indicate that seasonal changes in resistance may have a genetic component.

Organisms can evade a stress by entering a resistant life cycle stage, switching to another resource, or by moving away. There is evidence for genetic variation for evasion within and between populations, particularly in the case of life-history changes in insects.

Phenotypic plasticity encompasses short term reversible changes in stress responses, acclimation responses, and irreversible changes during development. Species and populations show different levels of plasticity. High levels of phenotypic plasticity for physiological traits occur in species from habitats experiencing environmental fluctuations, while levels of plasticity for morphological traits are relatively low in species from stressed environments.

Whether phenotypic variation for stress response traits within and between populations is associated with genetic variation or plasticity depends on a number of factors, including the frequency, magnitude and arrival speed of the stress periods, selection history, physiological constraints, costs of plasticity, and reliability of environmental cues.

5. Effects of stress on genetic variation

Environmental stresses act as selective agents that can alter gene frequencies in populations and so cause genetic divergence within and among populations, as shown by the examples in the previous chapter. Stress can also have more immediate effects on the genome and on the expression of genetic variation at the phenotypic level, and we consider these effects in this chapter. We examine the consequences of adverse conditions on the expression of genetic variation by reviewing data on stress and heterosis, and by examining data on the impact of environmental conditions on the genetic component of variance and the relative importance of the additive, dominance, and epistatic effects of this component. Explanations for changes in these genetic parameters with stress levels are discussed. Finally, we briefly consider the number of loci underlying genetic variation in stress response traits, as well as factors that may contribute to the persistence of genetic variation for these traits in populations.

5.1 Recombination, genomic stress, and mutation

Early evidence suggested that the incidence of recombination is lowest under optimal conditions and increases as the environment becomes progressively more stressful. During the first decade of genetic experiments on *D. melanogaster*, Plough (1917) found that recombination, especially in centromeric regions of chromosomes 2 and 3, increased when the temperature at which the female fly developed was increased or decreased from the normal culture temperature (25°C). For the black–purple region, just proximal to the centromere in chromosome 2, Plough found three fold increases in recombination at 13°C and 31°C, giving a U-shaped incidence curve (Fig. 5.1). Qualitatively similar results were obtained by Graubard (1932) over the narrower range of 14 to 30°C, with a two fold increase in recombination at the extremes. These temperature extremes correspond to marginal conditions for *D. melanogaster* because it is difficult to collect *D. melanogaster* below about 12°C in temperate zone habitats , and there is an upper threshold for species continuity in the range 30 to 31°C (Parsons 1977, 1983). Plough's (1917) data give a weighted recombination fraction of 8.7 per cent at 29°C and 18.2 per cent at 31°C, compared with a control value of

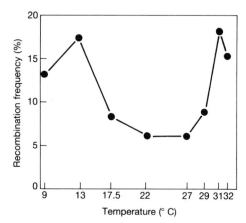

Fig. 5.1. Weighted recombination percentages for the black–purple region of chromosome 2 of *Drosophila melanogaster* plotted to show the effect of temperature on crossing over. (After Plough 1917.)

6.0 per cent, indicating the sensitivity of recombination to minor temperature changes under stressful conditions. There is, therefore a sharp increase in recombination at both high and low temperatures, close to environmental conditions that are marginal for *D. melanogaster*. Many additional experiments, such as those of Grell (1978), also indicate that recombination increases at temperatures above and below normal culture temperatures, especially in centromeric regions.

Temperature effects upon recombination have been studied in a number of other eukaryotic organisms, including the nematode *Caenorhabditis elegans*, the tomato and several fungi such as *Neurospora*, *Sordaria*, *Coprinus*, and *Schizophyllum* (reviewed in Parsons 1988). Many of these experiments provide suggestive evidence for higher recombination rates at temperature extremes, although generalizations are difficult because comprehensive experiments covering the entire temperature range over which fertile offspring can normally be produced have not usually been carried out.

Factors other than temperature may also influence recombination. In *D. melanogaster*, Neel (1941) found that recombination along chromosome 3 was increased by 0 to 15 per cent following starvation, while Aslanyan *et al.* (1987) described increases in recombination under the acceleration experienced during space flight. Recombination may also be affected by genomic changes, and there is a substantial literature on the interchromosomal control of recombination (Lucchesi and Suzuki 1968), indicating that structural heterozygosity in one part of the genome usually increases recombination in the remainder of the

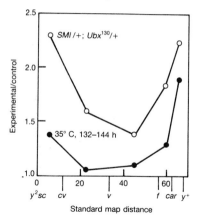

Fig. 5.2. Crossing over in five regions of the X chromosome of *Drosophila melanogaster* after exposure to heat stress (closed circles) and to heterozygous inversions in chromosomes 2 and 3 (open circles). The graphs give the relative change in crossing over in the X chromosome expressed as the ratio experimental/control. (Modified from Grell 1978.)

genome. In parallel with environmental variables, the major effect is in the region of the centromere with a lesser, but marked distal increase (Schultz and Redfield 1951). Strikingly similar patterns of response have been obtained for recombination (Fig. 5.2) when heat treatments (12 hours at 35°C) and interchromosomal effects were considered simultaneously (Grell 1978).

Several hypotheses have been put forward to explain interchromosomal effects on recombination (Lucchesi and Suzuki 1968). The parallel with temperature effects suggests that a common mechanism may be involved, although it is not known if various stresses have cumulative effects on recombination rates. Lucchesi and Suzuki (1968) suggested the involvement of enzymes that are operative in homologously paired regions at the time of crossing over. Increased recombination may occur when these enzymes do not function or attach normally, and this could account for changes in response to temperature as well as in response to non-homologous regions of the chromosome. Borodin (1987) found that immobilization, heat, or social interactions between male mice inhibited DNA synthesis in testes, and suggested that hormones released in response to stress induce an inhibition of replication and repair synthesis of DNA, which may disrupt synapsis, and influence recombination rates. Agents that inhibit DNA synthesis induce over-replication of DNA, and this is associated with changes in the levels of some proteins and the generation of chromosomal rearrangements and aberrations (Schimke *et al.* 1986).

Environmental stresses increase mutation rates, as well as recombination rates. The most obvious examples are chemical stresses that influence mutation rates directly, by acting as mutagens. There is also an enormous literature on the effects of ionizing radiation upon mutation (reviewed in Sankaranarayanan 1982). Considering other stresses, Schmalhausen (1949) cited increased mutation rates resulting from factors such as extreme temperatures, intensive isolation, extreme limits of humidity, chemical influence of salts, and unusual organic compounds in food. A survey of experiments on *Drosophila*, microorganisms, fungi and plants (Lindgren 1972) described 48 studies showing a positive relationship between mutation rate and temperature, with 12 studies showing no temperature effect and three studies showing a negative relationship. Lindgren's review also included five *Drosophila* studies which considered the effects of a cold shock treatment. Four of these studies gave an increase in mutation rate after a cold shock, suggesting that temperature extremes are important rather than high temperature *per se*.

Several mechanisms may account for the increased mutation rate of organisms placed under stressful conditions. One possibility is that stress triggers an error-prone repair pathway. This normally occurs when microorganisms are exposed to agents that damage DNA, such as UV light and many chemicals (Walker 1984), but error-prone pathways may also be induced in response to other stresses. For example, MacPhee (1985) found that spontaneous reversion rates of a *trp* gene were increased in *Salmonella typhimurium* grown on glycerol, rather than the more favourable carbon source, glucose. A mutation-generating function was switched on during growth on glycerol and the effect of this function was enhanced by exposure to a high temperature stress. The mobilization of transposable elements provides another mechanism for increasing mutation rates under stress. McClintock (1984) has argued for the importance of stress, both within the cell and imposed from without, as a trigger for rapid genomic reorganizations, some of which may occur via transposons, present in perhaps all prokaryotic and eukaryotic genomes. As early as 1951, McClintock reported an elevated mutation rate from the movement of mobile genetic elements in temperature-stressed maize plants, and there is evidence that transposition rates in maize, yeast and *Drosophila* increase with environmental temperature and with stress at the genomic level (Paquin and Williamson 1984; Strand and MacDonald 1985), although not all environmental stress experiments have given positive results (Parsons 1987*b*).

Walbot and Cullis (1985) have argued that environmental conditions can induce mutational changes in the plant genome which can be rapidly transmitted during both the mitotic and meiotic cycles. Many of these genomic changes may be translated into phenotypic variation. Following pioneering work of Durrant (1962), much of the research in this area is being carried out on flax, where heritable morphological changes occur after plants have been grown for one generation in certain controlled environments with imbalances in nutrients or temperature. Genomic changes in response to these treatments have been reviewed by Cullis (1987). The majority of the genomic changes involve alterations in the number of genes coding for rRNAs and in other highly repeated sequences, although it is not clear which of the changes is responsible for the phenotypic variation. The mechanism responsible for these changes is also unknown. It has been speculated that rapid genomic reorganization is important in evolutionary responses to environmental stress (Walbot and Cullis 1985), but there has been no evaluation of the relative importance of pre-existing genetic variation compared to variation generated *de novo* in response to selection for stress resistance. The importance of mutations in long term responses to selection is suggested by several artificial selection experiments for morphological traits in *Drosophila*. For example, Frankham (1980, 1988) found that mutations not present in unselected base populations contributed to responses to selection for abdominal bristle number in *D. melanogaster*. These mutations were generated by unequal crossing over.

Most mutational changes are expected to be maladaptive. New mutations often have large deleterious effects, as expected from the arguments of Fisher (1930) outlined in Section 1.2 (see Fig. 1.2). This makes it unlikely that individuals with relatively high mutation rates would be favoured under models of individual selection, except when the environment was continuously changing (Gillespie 1981) or in situations where only offspring outside the normal phenotypic range of the population survive. Cox and Gibson (1974) provided experimental evidence that high mutation rates in *Escherichia coli* can be favoured by natural selection, showing that genotypes with high intrinsic mutation rates were favoured in a novel chemostat environment where glucose was limiting and traits including starvation resistance were selected.

Theoretical models suggest that natural selection will often adjust mutation rates to intermediate levels, between very high levels that cause an increase in the mutational load and very low levels that do not allow adequate variation for populations to adapt to unpredictable changes in the environment (e.g. Kimura 1967; Gillespie 1981). This idea was substantiated experimentally by Nöthel (1987) in a study of *D. melanogaster* populations with irradiation histories extending over 600 generations. He found that the irradiated populations had genetic factors for radioresistance which reduced the frequency of germ cell mutations induced by irradiation. Removal of the irradiation treatment caused an increase in the mutation rate, while increasing the irradiation rate led to the evolution of additional factors decreasing irradiation damage. This suggests that mutation rates evolved in response to environmental conditions.

Although the response of mutation rates to stress have not been addressed theoretically, it seems likely that mechanisms increasing mutation rates under stress can be adaptive because they enable organisms to generate genetically variable offspring in response to environmental extremes when populations may face extinction (Wills 1983; Walbot and Cullis 1985). DNA repair mechanisms, such as the error-prone SOS pathway, may be triggered under stressful conditions and provide mutants capable of countering the injurious effects of an environmental change. However it is also possible that the increased rates of genomic change under stress are not adaptive. They may simply be a consequence of cellular changes induced by stress that increase susceptibility to DNA damage or decrease the ability of cells to repair such damage. Nevertheless, there is variability in the ability of species and individuals to undergo genomic change in response to stress (e.g. Durrant and Timmis 1973), suggesting the potential for natural selection to influence levels of mutation under stress.

Similar considerations apply to the increased recombination rate in response to stressful conditions. Recombination levels are influenced by genetic factors. For example, Charlesworth and Charlesworth (1985) found that levels of recombination in *D. melanogaster* were influenced by genes on all three major chromosomes. This genetic variation indicates that recombination rates can be altered by natural selection, which is expected to produce a tightening of linkage between genes affecting fitness in a homogeneous environment because the same gene combinations are always favoured (Fisher 1930). Conversely, a loosening of linkage may be selected in variable environments when different gene combinations are favoured under different conditions (Mather 1973; Charlesworth 1976). Zhuchenko *et al.* (1985) selected *D. melanogaster* populations for resistance to extreme temperature fluctuations and found a substantial increase in recombination in the centromeric region of chromosome 2 in the selected lines when tested under optimal conditions. Selection for resistance to variable temperature stress may therefore lead to the evolutionary alteration of recombination rates. Flexon and Rodell (1982) found similar results in *D. melanogaster* lines selected for increased resistance to DDT. Recombination rates increased in chromosome 2 and 3, and these were also the chromosomes where genes increasing DDT resistance were mapped. This suggests that genes influencing recombination rates were selected because recombination produced favourable combinations of genes that increased DDT resistance. Recombination levels may therefore be higher in organisms that are exposed to novel habitats where extremes of climate or other stress factors are encountered, or in organisms from habitats that experience fluctuating conditions.

These results do not directly address the adaptive significance of higher recombination rates under stressful conditions because changes in the response of recombination to environmental conditions were not scored. However experiments on tomatoes by Zhuchenko *et al.* (1986) indicate that individuals differ in the extent to which their recombination rates are influenced by environmental extremes. Reduced temperature did not increase the recombination index in two cold-resistant varieties and an F_1 hybrid, but did increase this index in a cold-susceptible variety. In contrast, high temperatures only increased the recombination index in cold-resistant forms. This suggests that the effects of environmental factors on recombination depend on the extent to which an individual is stressed. It is not known whether there is genetic variation for increased recombination rates under stress that is not simply a consequence of the stress experienced by an organism. If such variation does not exist, then increased recombination under stress may be a consequence of the stressed state rather than an adaptive response.

5.2 Canalization, asymmetry, and stress

Waddington (1956) studied the influence of stress on the expression of genetic variation in morphology, subjecting eggs of *D. melanogaster* to ether vapour in sub-lethal doses. Most of the survivors were normal, but a few developed an abnormal phenotype known as bithorax, where the thorax appears to be duplicated. Abnormal flies were used as parents in subsequent generations and the ether treatment was repeated on their eggs. The incidence of bithorax increased steadily in this selected line, and the line eventually produced the bithorax condition even without exposure to ether. Bithorax had become a genetically determined trait within 30 generations. Waddington referred to this process, where phenotypes produced by a stress become encoded by the genome, as 'genetic assimilation'. These experiments have been replicated for different traits produced by other severe environmental stresses. For example Waddington (1953) found that a short period of sub-lethal high temperature at a certain critical stage in pupae caused the appearance of a phenotype known as crossveinless (cv) where crossveins in wings are absent or reduced. The cv phenotype can also be produced by a mutant allele, but this allele was absent. By selecting those flies with cv, Waddington produced strains with a large number of crossveinless individuals which were not genetically homozygous for the mutant cv allele. Milkman (1960) also selected for crossveinless flies following temperature treatment and found that the production of cv in the absence of a stress was controlled by genes of substantial effect, present on the major chromosomes of natural populations of *D. melanogaster.* Various other abnormalities were produced and selected by Bateman (1959). In many cases a single chromosome was responsible, suggesting control by a major gene, but several genes were implicated in other cases.

Three points are evident from these examples. First, there is a certain specificity of the original stimulus; eggs exposed to ether vapour give bithorax, and pupae exposed to high temperatures give crossveinless. Secondly, the sub-lethal environments represent severe stresses, and greatly increase the variability of development. Thirdly, these severe stresses reveal much unexpressed genetic variation in the original population, and some of the unexpressed genes, singly or in combination, have a tendency to produce an abnormal phenotype such as bithorax. This variation may normally be suppressed by self-regulatory developmental processes, for which Waddington (1956) used the term canalization. In other words, a specific environmental stress augments variability by decreasing canalization and this results in the production of a weakly developed bithorax condition in individuals having a genetic tendency towards bithorax. Continued selection of such individuals leads to a rapid increase in genes producing the bithorax condition, until the specific environmental stimulus eventually becomes unnecessary.

Williams (1966) regarded these experiments as important in demonstrating a previously unsuspected store of genetic variability involving the epigenetic system, but did not consider them to be an important model of adaptive evolution. Nevertheless, these findings may be relevant to the extreme stress periods that are encountered by natural populations, and it is possible that reduced canalization may lead to the production of stress-adapted phenotypes under extreme circumstances. A limitation of these *Drosophila* experiments is that the types of traits that were examined are not directly involved in responses to stressful conditions, and the phenotypes are frequently so abnormal that they would be unlikely to survive in nature. A more relevant situation for natural populations is the effect of temperature variation (18–32°C) on the development of meristic characters in the snake, *Natrix fasciata* (Osgood 1978). The frequency of gross abnormalities in vertebrae was increased at both high and low temperatures. Studies on the thermoregulatory behaviour of gravid females indicated a preferred body temperature around 26.5°C, which corresponded closely with the minimum frequency of abnormal embryos, in the 25 to 27°C range. Extreme temperatures have also been associated with floral abnormalities in field populations of *Linanthus* (Huether 1969). These types of studies are logistically difficult and a trait that can be generalized more readily across taxa is needed.

From this point of view, one epigenetic process that has attracted attention over many years is asymmetry (Palmer and Strobeck 1986) which is shown by most bilateral characters. There are three kinds of asymmetry, which are referred to, in order of decreasing frequency of occurrence, as fluctuating asymmetry, directional asymmetry, and anti-symmetry (Van Valen 1962), and are characterized by differing combinations of right-minus-left differences $(R - L)$. In fluctuating asymmetry, there is a normal distribution of $R - L$ values around a mean of zero. In directional asymmetry, the mean measurements for a bilateral trait are almost always greater on one side than on the other side. In anti-symmetry, there is also a difference between the sides, but it cannot be predicted which side will show the greater value, giving a bimodal distribution of $R - L$ values about a mean of zero (Palmer and Strobeck 1986).

Fluctuating asymmetry has been studied for paired characters, including numbers of scales and chaetae, wing venation, and skeletal characters, in a wide variety of organisms. It has often been proposed that fluctuating asymmetry provides a measure of developmental stability (Mather 1953; Thoday 1956; Beardmore 1960; Reeve 1960; Parsons 1961, 1962; Van Valen 1962; Soulé 1967). As the level of stress of any kind increases, fluctuating asymmetry and its variability should increase, provided that the stress is sufficiently intense to influence developmental processes. Figure 5.3 gives results of an experiment by Parsons (1962) with an Oregon-R stock of *D. melanogaster* grown at 25°C and 30°C. There was far more fluctuating asymmetry of sternopleural chaeta number at 30°C over a range of maternal ages, and especially in flies derived from young females, when developmental instability appeared to be particularly high. At 30°C, there was a great deal of variability from day to day, which tended to obscure any association between fluctuating asymmetry and age. The variability at 30°C might be expected, since the extreme nature of this stress meant that slight changes in microenvironment may have had large effects. At 25°C, a linear regression indicated that fluctuating asymmetry increased with maternal age ($p < 0.001$), a result consistent with many other observations on increased developmental instability with maternal age, especially in mice and humans (Parsons 1964).

In a further experiment, sternopleural chaeta number and fluctuating asymmetry were higher at 30°C than at 25°C; in addition, the increased developmental instability at 30°C was confirmed by the common occurrence of flies with abnormalities including crumpled and curled wings (Parsons 1961). Fly weight was only slightly reduced at the higher

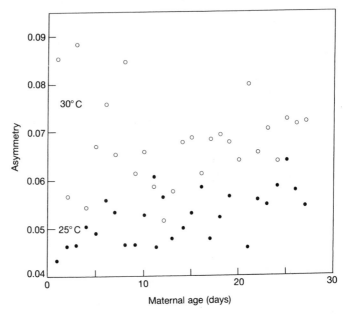

Fig. 5.3. Asymmetry of sternopleural bristle number plotted against maternal age for the Oregan-R stock grown at 25°C and 30°C. Each point represents the asymmetries for a group of 60 flies. Asymmetry was measured as the difference between left and right counts divided by the total bristle number. (After Parsons 1962.)

temperature, and emergence time was reduced. In contrast with high temperature stress, a chemical stress, phenylthiourea, substantially reduced fly weight and caused a delay in emergence time, as well as increasing the variability for these traits. However the chemical had little effect upon chaeta number variability and fluctuating asymmetry, suggesting that it does not influence the developmental stability of flies. Presumably, the contrast between the effects of these stresses resides in the fact that temperature is a generalized stress with multiple conse-quences for the basic physiology of the organism, whereas phenyl-thiourea is a specific chemical stress, affecting far fewer metabolic pathways than temperature.

An experiment with *D. melanogaster* demonstrated that levels of fluctuating asymmetry are under genetic control and may be reduced during adaptation to temperature stress (Thoday 1958). Sternopleural chaeta number and fluctuating asymmetry were studied over 30 gener-ations at 25°C, the normal culture temperature, and at three 'novel' environments, 20°C, 30°C and one changing gradually between 20°C and 30°C in a diurnal cycle. Genetic changes in response to the new environments were indicated by a change in chaeta number, suggesting that this measured something of adaptive significance. In particular, chaeta number increased at 20°C and decreased at 30°C, which is consistent with the well-known observation that flies tend to have fewer chaetae when cultured at high compared with low temperatures (Parsons 1961). The results also provided evidence that sternopleural fluctuating asymmetry is a trait of adaptive significance, since there was no indication that this changed consistently in the 25°C population, but each of the other populations in new environments showed a negative regression of fluctuating asymmetry on generations. The regression was significant in the 30°C and the 20/30°C populations, and close to significance in the 20°C population. The lower significance of the 20°C results may reflect the fact that this temperature is not a major environmental perturbation by comparison with 30°C, which is at the limits of survival for *D. melanogaster.*

Several other lines of evidence from *Drosophila* experiments indicate that fluctuating asymmetry is reduced during adaptation. Fluctuating asymmetry was normally reduced significantly in the F_1 and F_2 generations from inbred strains (Mather 1953; Reeve 1960), consistent with its role as a fitness measure. Inbreds tend to have inferior fitness compared to the hybrid generation as measured by other traits such as fertility and viability. Reeve (1960) found increased fluctuating asymmetry in flies with major mutations in the chaeta system. Major mutations tend to have deleterious fitness effects, suggesting that developmental homoeostasis is reduced in the mutant genetic background. Similar data have been obtained in the sheep blowfly, *Lucilia cuprina,* where initial asymmetry associated with a major gene providing resistance to the insecticide diazinon was reduced by the subsequent evolution of modifiers (McKenzie and Clarke 1988).

There is a substantial literature on fluctuating asymmetry in organisms other than insects (Palmer and Strobeck 1986) which provides parallel conclusions. Considering the dentition of house mice under controlled conditions, as expected:

(1) hybrids showed less FA than inbreds (Leamy 1984);

(2) directional selection for body weight led to increased FA (Leamy and Atchley 1985);

(3) various stresses, in particular cold, audiogenic stress, protein deprivation and heat increased FA (Siegel *et al.* 1975; Sciulli *et al.* 1979).

In fish, Valentine and Soulé (1973) increased pectoral ray asymmetry in fry of the Californian grunion, *Leuresthes tenuis,* by exposure to DDT.

The experimental data suggest that environmental effects on fluctuating asymmetry will be difficult to detect in the field, since conditions of relatively severe stress are needed. One approach is to seek out eco-logically marginal situations where stressful conditions are most likely to occur. For example, Soulé and Baker (1968) scored a set of six characters in Rocky Mountain populations of the butterfly, *Caeno-nympha tullia*, and found that the most asymmetrical populations occurred at high altitudes which are subjected to climatic extremes. In three species of fish, fluctuating asymmetry showed an increase approaching heavily populated areas from the north and the south in southern California (Valentine *et al.* 1973; Valentine and Soulé 1973). The tentative explanation for the increase was exposure to environ-mental contaminants, suggesting that fluctuating asymmetry may be a useful monitor of stresses such as pollution. In the muskrat, *Ondatra zibethicus*, Pankakoski (1985) used the speed of individual growth as a measure of habitat suitability. Foramen numbers on both sides of the skull were counted in the same animals to test the hypothesis that there is a negative relationship between habitat suitability and degree of mor-phological asymmetry of skull nerve foramina. The correlation between asymmetry and the growth index was -0.94 ($p < 0.001$), indicating a strong relationship between population fluctuating asymmetry means and the suitability of habitat as reflected by growth indices. These studies confirm the merit of looking at different populations from diverse environments, and are consistent with experiments under controlled laboratory conditions, in showing that there is a tendency for fluctuating asymmetry to increase as environments become increasingly stressed. There are also many inconclusive reports in the literature which may be explained in terms of inadequate sample size or insuf-ficiently severe stress levels (see Parsons 1990 for discussion).

In summary, the degree of fluctuating asymmetry in many traits can generally be related to environmental stress. The expectation is for increasing fluctuating asymmetry away from an optimum, especially at stress levels approaching lethality. Genetic changes reducing fluctuating asymmetry can occur when organisms are held in new environments, as shown by a progressive fall over a number of generations in *Drosophila* exposed to stressful conditions. There appears to be a growing emphasis upon the use of fluctuating asymmetry for measuring stress levels in natural populations (Leary and Allendorf 1989). However the effects of variation in stress intensity on this parameter in natural populations will need to be considered in assessing its usefulness.

5.3 Heterosis and stress levels

Heterosis has been defined as

the increased vigour of the F_1 over the mean of the parents or as

the increased vigour of the F_1 over the better parent
(Sedcole 1981), although it is often only used in the second context.
Hybrids are distinguished from parental strains by being heterozygous at
more loci. Heterosis is usually characterized in crosses between strains
that differ at a number of loci, although it is also used in the context of
inversions that lock up groups of genes. Heterosis may also refer to the
heterozygous genotype of a single locus although the term 'heterozygote
advantage' is commonly used in this situation.

A tendency for the difference between parents and F_1 hybrid genotypes to increase under stress has been found in a wide range of studies.
Starting with chromosomal inversions in *D. pseudoobscura*, Wright and
Dobzhansky (1946) demonstrated that inversion frequencies evolved
rapidly to a stable equilibrium in population cages kept at 25°C, and that
this equilibrium reflected an advantage of inversion heterozygotes over
inversion homozygotes. This temperature is extreme for *D. pseudoobscura* in comparison with 16.5°C, where only small fitness differences
among inversion types were found because inversion frequencies did not
change much in population cages. An intermediate situation occurred at
22.5°C (Van Valen *et al.* 1962), suggesting that the degree of heterozygote advantage increased with temperature stress.

Similar findings have been made in crosses between strains of
Drosophila and in the plant, *Arabidopsis*; hybrids derived from crosses
between strains tend to perform better under high temperatures than the
parental strains, while this difference is less evident at intermediate
temperatures (Langridge 1962; Parsons 1971). Differences between
strains and hybrids do not always increase under extremes of low
temperature, although increased heterosis under stress has also been
found with other types of stressors.

These findings can be illustrated by an extensive experiment on hybrid
superiority in *Arabidopsis thaliana* carried out by Pederson (1968).
Four environmental factors (temperature, light, available nutrients and
water) were varied to give reductions in growth rate of 25 to 50 per cent.
The growth of 10 geographic strains, 15 F_2 hybrids and 15 F_2 hybrids
from crosses of different F_1s (that is, double-cross hybrid) were compared. The strains were considered largely homozygous because they
had been maintained by self-fertilization. The degree of heterozygote
advantage for growth increased with stress for all environmental variables except nutrient concentration. This was evident in the parental
strain–F_2 comparison and the comparison between hybrids and double
hybrids, particularly at high stress levels (Table 5.1).

The literature on hybrid advantage and stress in agricultural animals

Table 5.1. Effects of environmental stresses on hybrid advantage for growth rate in *Arabidopsis thaliana* measured as the weight of plants after 14 days. Weights were log transformed and used to calculate the difference in biomass between the F_2 and mid-parent value, and the difference between F_2s derived from crossing the F_1s of the same or different parents. (After Pederson 1968.)

Environmental factor	Level of stress	F_2 – mid-parent	Double hybrid – F_2
Temperature	Absent	0.067*	0.028
	Medium	0.176	0.109*
	High	0.246‡	0.386‡
Light intensity	Absent	0.067*	0.028
	Medium	0.050*	0.028
	High	0.084*	0.167‡
Mannitol concentration	Absent	0.067	0.028
	Medium	0.033	0.050†
	High	0.120†	0.157‡
Nutrient concentration	Absent	0.067*	0.028
	Medium	0.033	−0.016
	High	0.054*	0.032*

*, † and ‡ indicate significant differences from zero at the 5%, 1% and 0.1% levels respectively.

has been reviewed by Barlow (1981). Heterosis generally varies with environmental conditions, and hybrids usually perform relatively better than parental strains under stressed conditions, particularly for sexual maturity and fertility. More variable results have been obtained for fecundity and growth rate, particularly under conditions of nutrient stress. For example, in ruminants heterosis for growth increases under stressful temperatures but decreases with nutritional stress. However the increase in heterosis with decreasing nutritional stress may not extend to other organisms. Recent experiments on larval weight in flour beetles, and on egg hatchability and production in *Drosophila* (Rich and Bell 1980; Ruban *et al.* 1988) indicated that heterosis increased as the level of nutritional stress increased. Changes in heterosis with stress may therefore depend to some extent on the type of trait and its association with fitness, as well as the nature of the environmental stress, and perhaps the degree of domestication of the species under investigation.

The above results have been obtained with crosses between inbred lines or crosses between parental strains originating from different populations where the strains have not experienced a history of inbreeding. Does heterosis increase with stress levels when genotypes are defined in terms of their heterozygosity and come from the same population? Relevant data are scarce, but suggestive evidence comes from the relationship between heterozygosity at allozyme loci and growth rate in individuals from the same population (Mitton and Grant 1984). Some findings (discussed in Section 7.5) suggest that this relationship is more likely to be detected in natural populations exposed to abiotic stress factors than in an optimal laboratory environment. However, the relationship between allozyme heterozygosity and fitness has not been systematically studied in a range of environments encompassing stressful and optimal conditions. Heterosis within populations is expected to change the additive and dominance components of variance as stress levels are increased. Experimental data, discussed below, suggest that genetic variance for stress resistance traits is often largely additive, which would suggest that the increased effects of heterosis under stress may not be important within populations.

Four explanations have been proposed to account for changes in heterosis with stress. The first is that parental lines are highly inbred, and heterosis largely represents a recovery from inbreeding depression. Heterosis will increase with stress if inbreeding depression is more evident in adverse environments than in favourable ones. Experimental verification would involve comparisons of hybrid and parental lines, originating from the same population but subjected to different degrees of inbreeding. This explanation does not account for heterosis in the many crosses where parental strains are not inbred, such as the comparisons of inversion homozygotes and heterozygotes.

A second possibility (Knight 1973) is that parental lines reach their optimum performance at different levels on a single environmental variable or a combination of variables. Hybrid genotypes are assumed to have an intermediate optimum between the two parents (Fig. 5.4a). The hybrid will then perform better than or equal to the mid-parental value across most environments, and there will be heterosis by environment interactions because the hybrid will outperform its parents at some levels of the environmental variable. This model is applicable to crosses between strains that are not inbred because the parents do not have a lower overall performance than the hybrid. However one environmental variable is not sufficient to generate increased heterosis in stressed environments. Knight illustrated how this could occur by considering different levels of a second environmental variable which interacts with the first variable (Fig. 5.4b). Taking sections through Fig. 5.4b at different levels of the first and second environmental variable can result in

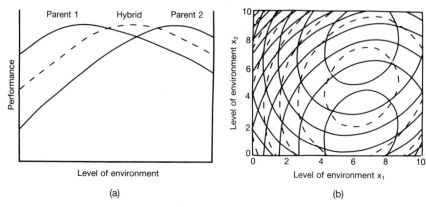

Fig. 5.4. Performances of 2 parental genotypes (solid lines) and their hybrid (broken lines) at different levels of (a) a single environmental factor where the hybrid is intermediate, and (b) two environmental factors where performance of the genotypes are indicated by the contours. (Adapted from Knight 1973.)

heterosis under extreme conditions (Fig. 5.5c), or heterosis under optimum conditions (Fig. 5.5a), or cases where hybrid performance increases or decreases relative to the mid-parental value (Fig. 5.5b,d). This model does not generally predict heterosis under stress, but may help to explain the contradictory results for nutritional stress. Knight's model has not been applied to the question of heterosis at individual loci, although it could be extended to enzyme variation if allozymes had different optima in relation to an environmental variable, heterozygotes achieved optimum activity at intermediate levels, and enzyme activity was directly related to fitness. We have already discussed examples where the activities of enzymes from different populations change rank as levels of an environmental variable are altered (Section 3.1), but the association between enzyme activity and fitness may be more complex, as discussed below.

A third explanation is that heterozygous genotypes have an advantage under stressful conditions but not in favourable environments. Heterozygote advantage may therefore arise because these genotypes are more successful at countering fluctuating conditions (i.e. they show greater homoeostasis), as proposed by Lerner (1954). Alternatively, heterozygotes may have an inherent advantage under stress, but evidence for this is limited in the case of allozyme loci (Clarke 1979). There is some theoretical justification for the homoeostasis hypothesis since heterozygous genotypes are expected to have a lower variance in fitness under variable environments, even when alleles have additive effects in a particular environment (Gillespie and Turelli 1989; Section 5.7). This

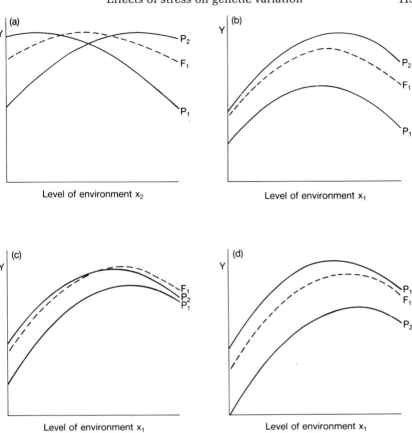

Fig. 5.5. Sections through the response surfaces from Fig. 5.4 when one of the environmental variables is held constant: (a) environment $X_1 = 7.00$; (b) environment $X_2 = 1.00$; (c) environment $X_2 = 6.00$; (d) environment $X_2 = 8.00$. (Simplified from Knight 1973.)

explanation may be relevant to some agricultural studies where stressful conditions are identified in terms of decreased productivity rather than controlled experimental conditions. Environmental variability will probably tend to increase as conditions for agricultural productivity deteriorate (Blum 1988). However this explanation does not account for cases of heterosis under stress when only one environmental factor is varied in controlled conditions.

Finally, Langridge (1962, 1968) proposed an explanation based on the premise that heat-sensitive enzymes are the most common consequences of mutations that do not inactivate the enzyme, and that some of these mutations are expressed only at high temperatures with complete dominance in the heterozygote. Heterosis under extreme temperatures could be based on these mutations if strains were fixed for different alleles. The deleterious effects of the temperature-sensitive alleles would be expressed in the homozygous strains but not in the heterozygous strain. Langridge provided evidence for such alleles at the β-galactosidase locus in *E. coli*. Numerous recessive temperature-sensitive genes are also known in *Drosophila melanogaster* and other organisms. This explanation is only applicable to crosses between inbred strains that are fixed for different deleterious alleles.

It is possible to extend Langridge's model to genes controlling variation in enzyme activity in natural populations. *E. coli* with genetic variation at structural loci for 6-phosphogluconate dehydrogenase (*Gnd*) and glucose phosphate isomerase (*Gpi*) had the same relative fitness in a chemostat (Dykhuizen and Hartl 1980, 1983). However differences at the *Gnd* locus were detected when an alternative metabolic pathway for 6-phosphogluconate was lacking and when cell density was high, while differences at the *Gpi* locus were detected when fructose was substituted for glucose. This suggests that allozyme variants may differ in fitness under some environmental and metabolic conditions. Similar results have been reported for variants at the locus for glucose-6-phosphate dehydrogenase in *D. melanogaster* (Eanes 1984): differences between variants were only detectable when the genetic background included a low activity mutant for a related enzyme (6-phosphogluconate dehydrogenase). Hartl *et al.* (1985) suggest that allozyme variants may only have demonstrable differences in fitness under 'stressful' conditions, although most of the 'stresses' that have been considered so far involve internal biochemical alterations rather than external environmental variables.

Results obtained in *E. coli* cannot be directly related to the performance of heterozygous and homozygous genotypes at enzyme loci, but it is likely that the performance of the heterozygote will be close to that of the fittest homozygote in a diploid organism. This follows from models of the relationship between metabolic flux and enzyme activity at steady-state conditions outlined by Kacser and Burns (1973, 1981). The effect of a change in activity of an individual enzyme on overall flux or reaction velocity through a metabolic pathway can be defined by its sensitivity coefficient, defined as the fractional change in flux over the fractional change in enzyme activity. The sum of the coefficients must be one in segments of complex pathways (Kacser and Burns 1981), which means that individual coefficients must be less than unity. This leads to the general relationship between activity and flux presented in Fig. 5.6. The non-linear nature of this relationship implies that the effect of

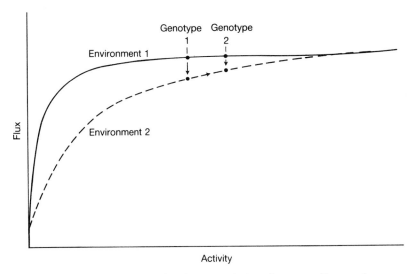

Fig. 5.6. Possible differences in the association between flux and enzyme activity in a favourable (environment 1) and a stressed situation (environment 2). (After Hartl *et al.* 1985.)

activity on flux is relatively greater at lower enzyme activities. Equating flux with fitness provides the basis for concave fitness functions (Gillespie 1976), and results in dominance for alleles with higher activity (Kascer and Burns 1981). The activity of wild-type alleles will be near the plateau of the curve under favourable conditions, and fitness differences between genotypes may not be evident. Hartl *et al.* (1985) have argued that fitness functions may change in an unfavourable environment so that genotypes which appeared to be neutral because they were at the plateau area of the flux curve now become associated with fitness differences (Fig. 5.6). These curve changes occur when the sensitivity coefficient for the enzyme under investigation is increased, so that the relationship between flux and activity becomes more linear, and differences between genotypes are therefore more likely to be detected under stressful conditions.

The above considerations may provide an explanation for increased heterosis under stress as an extension of Langridge's model. Two lines that are homozygous for a number of loci may both be homozygous for some alleles that reduce flux under extreme conditions, but not under favourable conditions. When these lines are crossed and tested under stress, heterozygous genotypes will tend to have flux similar to that of the favoured homozygote, resulting in heterosis in the stressed environment when averaged over a number of loci. Differences between hetero-zygotes and homozygotes will be accentuated in inbred lines when alleles having large deleterious effects on fitness (large decreases in enzyme activity) have been fixed. Such alleles can reduce flux considerably and will always be recessive (Kacser and Burns 1981). Crosses between inbred strains that are each fixed for several different deleterious alleles should produce a substantial increase in heterosis under stress. Crosses between populations are also more likely to produce heterosis under stress than crosses within populations because populations will differ for a larger number of alleles. It should be emphasized that this expla-nation does not assume heterozygote advantage at individual loci, but only requires that alleles associated with high flux are present in both of the lines that are crossed. In this sense, it is analogous to the 'dominance hypothesis' for heterosis (see Sedcole 1981), where the dominance relationship between the alleles rather than heterozygote advantage explains heterotic effects.

The association between activity and flux may also provide an explanation for the association between fitness and multilocus hetero-zygosity, discussed further in Section 7.5. Heterozygotes will not be fitter than homozygous genotypes when only one locus is considered, but this will change when heterozygosity is considered over many loci because the fitness of the heterozygote at each locus will be closer to that of the fittest homozygote. Individuals that are heterozygous at several loci could therefore have a higher mean fitness than homozygous individuals, even though some of the homozygous individuals will have the same fitness or higher fitness than the heterozygotes.

There are many situations where fitness under stressful conditions will not be positively associated with flux. As suggested by Beaumont (1988), the expression of many traits under stabilizing selection, as opposed to directional selection, may depend on the size of substrate pools that represent intermediate steps in metabolic chains rather than on overall flux through a pathway. Another complication is that bio-chemical pathways are interlinked, so that high levels of flux through one pathway may adversely affect a fitness trait associated with flux through a linked pathway. Finally, the association between flux and enzyme activity outlined in Fig. 5.6 depends on steady-state dynamics which may not hold under all stressful conditions.

In summary, increased heterosis under stress is common, and may be explained by several different models. Some of these models predict heterosis even when parental strains are not inbred and fixed for deleterious alleles. The interaction between heterosis and environ-mental stress does not necessitate an inherent advantage for hetero-zygous genotypes.

5.4 Changes in heritability with stress: empirical data

The phenotypic variance (V_P) of a trait is normally expressed in quantitative genetics by the equation

$$V_P = V_A + V_D + V_I + V_E.$$

where V_A is the additive genetic variance, V_D the dominance variance, V_I the variance due to epistatic interactions between genes, and V_E the environmental variance. Changes in the narrow heritability (V_A/V_P) or broad heritability ($V_A + V_D + V_I/V_P$) can be caused by changing any of the three genetic components or the environmental component.

Numerous experiments have demonstrated that the magnitude of the genetic and environmental components of variance changes with environmental conditions, and these findings have often been interpreted in terms of changes in heritability with stress levels. Two opposing views have emerged: one is that extremely stressful conditions increase heritability, particularly the narrow heritability (Parsons 1974, 1983); the other view is that stressful conditions lead to a decrease in the heritability (Johnson and Frey 1967; Blum 1988). As will be shown below, partial reconciliation of these views may be reached by considering the empirical data on which they are based. These data encompass a wide variety of stress definitions, experimental techniques, and experimental organisms. Explanations for the changes in genetic parameters with stress levels are examined in the next section.

5.4.1 Animal studies

The type of experiment where heritability has been found to increase with levels of stress is illustrated by a comparison of *D. melanogaster* isofemale lines for adult longevity during exposure to different concentrations of chemicals (Parsons 1983). In this species, vapours from 3 per cent ethanol and 6 per cent acetic acid are used as resources by adults, since exposure to these chemicals increased longevity substantially, whereas 12 per cent ethanol and 12 per cent acetic acid are stressful, since exposure to these concentrations shortened longevity (Parsons 1982b). Table 5.2 shows that intraclass correlations for longevity comparisons of isofemale lines from different Australian populations were >0.5 for the stressful concentrations and <0.5 for the optimal concentrations. The two highest correlations were for the Townsville population, which also had the shortest longevities and was, therefore, most susceptible to these stresses. The intraclass correlations reflect the size of the genetic component of variance relative to the environmental component, although they are not equivalent to heritability in this particular example because measurements were made on

Table 5.2. Intraclass correlations from analyses of variance within and between nine isofemale strains of *Drosophila melanogaster* for Melbourne, Brisbane and Townsville populations scored for resistance to ethanol and acetic acid vapour. (Data from Parsons 1982*b*.)

	Ethanol		Acetic acid	
	3%	12%	3%	12%
Melbourne	0.38	0.68	0.33	0.68
Brisbane	0.35	0.56	0.35	0.76
Townsville	0.16	0.82	0.46	0.87

groups of flies rather than on individuals. The genetic component of the phenotypic variance therefore increased relative to the environmental component with stress levels in each population, and was also higher in the population where equivalent chemical concentrations were perceived as a relatively greater stress. The environmental stress in this experiment was clearly defined as different concentrations of a chemical, and the stressor was imposed on adults after they had been cultured under optimal laboratory conditions.

Another example of this approach is the study of Westerman and Parsons (1973) who carried out diallel crosses among four inbred strains of *D. melanogaster* and exposed the adult progeny to different levels of γ-irradiation. Resistance was measured in terms of longevity, and the diallel results were used to obtain the general combining ability, a measure of additive genetic effects, and the specific combining ability, a measure of mainly dominance effects. At the extreme dose of 120 Krad, the dominance component was small but significant, and the additive component predominated (Table 5.3). In contrast, the non-additive variance was more important at lower doses. The error mean square which reflects the environmental variance decreased with increasing dose. The narrow heritability therefore increased as the stress became more severe, while the dominance component showed the same trend as the error mean square. Dominance was in the direction of increased resistance. This experiment used a well-defined stress and adults were cultured in a favourable environment before they were stressed.

Genetic variation for the resistance of *Drosophila* to other environmental stresses may also be mainly under additive control. A 4×4 diallel involving two *D. melanogaster* strains sensitive to phenylthiourea and two resistant strains indicated that strain differences

Table 5.3. Results from diallel crosses between four *D. melanogaster* inbred strains scored for longevity after exposure to ^{60}Co-γ-irradiation between 0 and 120 Krad. Mean squares for general combining ability (gca) measuring additive effects and specific combining ability (sca) measuring nonadditive effects are given. (Modified from Westerman and Parsons 1973.)

	Dose (Krad)					
	0	40	60	80	100	120
Mean longevity (days)						
Males inbred	32.9	20.1	14.0	11.0	7.7	2.2
F_1s	40.6	24.8	16.8	13.6	10.3	3.1
Females inbred	29.2	21.0	18.6	14.9	9.6	3.9
F_1s	41.6	35.9	28.0	22.1	15.6	6.7
Analysis (mean squares)						
gca	669.7†	122.0	54.4	14.1	24.6*	143.7‡
sca	380.5*	221.9‡	83.6†	52.0‡	43.7‡	19.6‡
Error	150.6	57.4	24.2	12.0	8.1	5.2

$*p < 0.05$; $†p < 0.01$; $‡p < 0.001$.

were mainly due to additive effects (Deery and Parsons 1972*a*). Diallel crosses between several independent sets of isofemale strains of *D. simulans* from Melbourne provided evidence that strain differences for desiccation resistance were almost entirely additive (McKenzie and Parsons 1974*b*). However crosses with *D. simulans* strains from Brisbane indicated significant dominance as well as additive effects; alleles controlling high stress resistance were dominant. There was no evidence for dominance in *D. melanogaster* lines selected for increased desiccation resistance which showed heritabilities in the range 0.59 to 0.69 (Hoffmann and Parsons 1989*a,b*). Differences among isofemale strains for ether resistance were mainly due to additive genetic effects, with no significant dominance component for males; a small significant dominance component for the females arose from directional dominance for increased stress resistance (Deery and Parsons 1972*b*). Finally, genetic divergence in *D. melanogaster* lines selected for resistance to knockdown by ethanol fumes was entirely due to additive effects (Cohan *et al.* 1989).

Flies in all of the above studies were cultured under non-stressful conditions before testing for resistance, with the exception of the phenylthiourea experiments, when larval emergence was used as a measure of stress. This experimental design has generally not been extended to other species. An exception is Derr's (1980) study of the American cotton stainer bug, *Dysdercus bimaculatus*. The heritability for timing of the first clutch in this bug, calculated from parent–offspring regression, increased from close to 0 under benign conditions to 0.3 to 0.4 under conditions of desiccation stress of the intensity encountered in nature. Egg clutches that provided the adult and progeny generations in this experiment were reared under non-stressful conditions.

Experiments where organisms are cultured under stressful conditions before being tested under stressful conditions also provide some evidence for an increase in heritability with stress levels. Baker and Cockrem (1970) selected for body weight in mice that developed in a hot (31°C), medium (21°C), or cold (7°C) environment. Body weight was about 15 per cent lower in the hot environment than in the other two environments, suggesting that the hot environment represented a moderate stress. Realized heritabilities based on four generations of selection were greater in the hot environment than in the other two environments, reflecting an increase in additive variance with stress. Heritabilities for males were 0.67 ± 0.06 in the hot environment and 0.41 ± 0.05 in the intermediate environment, while comparable values for the females were 0.60 ± 0.08 and 0.33 ± 0.22 respectively. Falconer and Latyszewski (1952) selected mice under restricted feeding or with food *ad libitum*; the former treatment reduced weight by about 12 per cent, and realized heritability was larger when mice were selected under restricted feeding.

Belyaev and Borodin (1982) carried out a series of experiments with mice, and interpreted their results in terms of stress increasing the genetic component of variance. Several traits were studied in offspring from a diallel cross involving three inbred strains which had been kept in two types of cages, a normal cage and a smaller cage where crowding occurred. Fourteen per cent of inseminated females in the normal cage did not produce any offspring, while the equivalent figure for crowded individuals was 29.4 per cent, indicating that crowding reduced reproductive fitness. Traits showed increased additive genetic variance under crowding, especially in the case of preimplantation mortality, litter size, and relative adrenal weight, while the environmental variance showed relatively smaller increases with crowding (Table 5.4). Non-additive effects for preimplantation mortality and litter size also increased under crowding, partly reflecting the expression of heterosis. These experiments did not involve a physical stress, although crowding may have indirectly induced a stress. The stress concept used by Belyaev and Borodin in interpreting these results is related to that developed by Selye (1950, 1956), discussed in Section 1.1.

'Stress' due to crowding has also been implicated as an explanation for

Table 5.4. Results from a diallel cross between three strains of mice held in uncrowded and crowded (stressed) populations. Mean squares are given for the general combining ability (gca) which is a measure of additive genetic variability and specific combining ability (sca) which measures non-additive effects. (After Belyaev and Borodin 1982.)

	Preimplantation mortality	Litter size	Adrenal weight	Thymus weight
Control population (uncrowded)				
gca	1.1	0.4	6.7	8432†
sca	1.2	4.4†	3.8	878
Error	0.7	0.5	1.9	1543
Stressed population (crowded)				
gca	2.6*	3.2†	21.7†	7200†
sca	4.2*	6.7†	1.7	650
Error	0.4	0.5	0.9	345

$*p < 0.05$; $†p < 0.01$.

the contradictory results that have been obtained with selection experiments for increased longevity in *D. melanogaster*. Lints and Hoste (1974) and Lints *et al.* (1979) found no consistent response to selection for longevity, while selection responses occurred in other experiments (Rose and Charlesworth 1981; Rose 1984; Luckinbill *et al.* 1984, 1989). Luckinbill and Clare (1985) and Clare and Luckinbill (1985) found that populations with high or uncontrolled numbers of competing larvae responded strongly to selection, but there was little response when larval density was low. Larval crowding decreased adult body size and increased development time, indicating suboptimal conditions. The expressed genetic variation in adult longevity may therefore be higher under 'stressed' larval conditions (Arking and Clare 1986).

In contrast to the above results, studies with agricultural animals have been interpreted as indicating that heritability decreases with increasing stress. For example, several studies have shown that the heritability of milk and butterfat yield in dairy cattle is larger at higher levels of production than at lower levels of production (reviewed in Pani and Lasley 1972). Production levels are assumed to reflect the degree of stress in the environment without identifying features responsible for reduced production. Although these findings appear to contradict those discussed above, they may not be relevant to experiments under more defined conditions for several reasons. First, mean yield differences between environments are often small (e.g. Van Vleck 1963) and a slight decrease in production does not correspond to stressful conditions as defined in Section 1.1, although this criticism also applies to some of the laboratory selection experiments mentioned above where weight reductions were relatively small. Secondly, lower production may not be a consequence of continuous suboptimal levels of the same environmental factor, but may reflect deviations from optimal conditions in several environmental factors. Variable environmental conditions may select for different genes at different times because of genotype–environment interactions (Chapters 6, 7). This will reduce the correlation between parents and their offspring and heritability estimates based on this correlation when parents experience a different set of conditions to their offspring. This will also reduce the response to selection under variable conditions because the same genotypes are not continuously selected. Thirdly, these results may not extrapolate to natural populations because agricultural animals are domesticated, and have been selected for high yield in a relatively favourable environment. Genes influencing stress resistance are likely to be at a different frequency in natural populations. This criticism is also relevant to experiments with *Drosophila* stocks that have been maintained in the laboratory for a long time.

There is, nevertheless, some evidence from experiments under more controlled conditions that heritabilities can decrease or remain unchanged as stress increases. Korkman (1961) selected for weight in mice under a high plane of nutrition and a low plane that reduced body weight by about 23 per cent. He found a higher realized heritability for mice selected in the high plane of nutrition, in contradiction to the results of Falconer and Latyszewski (1952) mentioned above. Genetic parameters for body weight in *Tribolium castaneum* were estimated from parent–offspring regression at two levels of nutrition: the poor environment produced individuals with half the body weight of those from the good environment (Hardin and Bell 1967). Narrow heritabilities were similar in the two environments (0.21 ± 0.06 and 0.19 ± 0.05 for good and poor environments, respectively), but broad heritabilities were larger in the good (0.97 ± 0.07) than the poor (0.69 ± 0.07) environment, indicating that a large fraction of the genetic variance was due to dominance, and that this fraction was inflated when conditions were less stressful. Using a sib analysis, Murphy *et al.* (1983) examined genetic variation for several life-history traits in *Drosophila simulans* at four temperatures, ranging from optimal to hot. There was no consistent increase in heritability at the highest temperature, and one of the traits showed a reduction in heritability. Other studies where heritability did not change with environmental conditions are reviewed in Pani and Lasley (1972).

Some overall trends are suggested by these findings. First, narrow heritability for survival and some other traits increases with environmental stress when individuals are cultured under non-stressful conditions and tested at different stress levels. This finding does not simply reflect changes in the environmental variance component, but needs to be extended beyond *Drosophila*. Secondly, crowding may increase the additive genetic variance for fitness-related traits. Thirdly, generalizations about narrow heritability may be difficult to make when organisms are cultured and tested under stressful conditions. This conclusion is largely based on nutritional stress and needs to be extended to other types of stress. Finally, the dominance component of variance may increase when crosses are made between inbred lines or populations because of heterosis (discussed in Section 5.3), but consistent changes may not occur in other cases. The effect of heterosis on the dominance component is evident in the crosses between inbred mice (Belyaev and Borodin 1982) and in *Drosophila* (Westerman and Parsons 1973). It should also be mentioned that there is little evidence for strong epistasis in those *Drosophila* experiments incorporating a design that could detect such effects (e.g. Parsons *et al.* 1969; McKenzie and Parsons 1974*b*; Cohan *et al.* 1989; Hoffmann and Parsons 1989*b*).

5.4.2 *Plant studies*

The association between heritability and stress in crop plants has attracted considerable attention because of the implications for plant breeding experiments aimed at obtaining rapid selection responses. Relevant studies have usually considered size and yield in individuals cultured under different environmental conditions.

Several experiments have provided evidence that genetic variance decreases with increasing stress levels (Blum 1988). For example, Johnson and Frey (1967) compared the growth of 27 oat cultivars under different nutritional conditions and different planting dates. These treatments affected several measures of plant performance. Genotypic variances generally increased as environments became more favourable, although environmental variances also increased so that heritabilities were not always higher under more favourable conditions. Similarly, Rumbaugh *et al.* (1984) found that genetic variance and broad sense heritabilities for shoot dry weight in alfalfa and wheatgrass seedlings declined as the amount of supplemental water was reduced. Conditions were varied so that there was almost no growth in the most extreme environment (representing a severe stress).

In other experiments, intermediate stress levels have been associated with the highest heritability, or heritability has not been influenced by the environment. Daday *et al.* (1973) grew 18 full-sib families of alfalfa

(*Medicago sativa*) in four Australian localities under three environmental conditions (high temperature, low temperature and high temperature combined with drought) at each locality. These conditions had large effects on growth rate, plant height in the poorest environment being only 10 per cent of that in the most favourable environment. Estimates of narrow heritability were 0.16, 0.22, 0.27 and 0.39, respectively, for plots from Narrabri, Glebe, Yanco and Charnwood under the most favourable conditions. The equivalent heritabilities for these locations under intermediate conditions were 0.27, 0.40, 0.46 and 0.38, respectively, indicating that heritabilities tended to be higher under intermediate conditions. Heritability estimates were close to zero under the most stressful conditions. Allen *et al.* (1978) considered yield in five crops (barley, wheat, oats, beans, flax) grown in favourable, unfavourable and intermediate locations for growth. Yield was 45 to 71 per cent lower in the unfavourable environments. Both the genetic and environmental variances were higher in the favourable environment, and heritability differences between environments tended to be small and were inconsistent in direction.

Finally, a few studies have recorded an increase in heritability with stress. Paroda and Hayes (1971) crossed 10 strains of barley and examined the rate of ear emergence in the parental lines and the F_1s in eight environments. Plotting the time of ear emergence against genetic components of variance indicated a positive relationship with the additive component but not the dominance component. Ear emergence, used as a measure of stress, was related to decreasing environmental temperature. Additive genetic variance therefore increased as environments became less favourable, while dominance variance did not change. Langridge and Griffing (1959) considered fresh weight in 43 homozygous races of *Arabidopsis thaliana* and obtained genotypic variances of 0.0128, 0.0115, and 0.0592 at 25°C, 30°C, and the stressful temperature of 31.5°C, respectively. Environmental variances showed smaller increases under stress, varying from 0.0407 at 25°C to 0.0711 at 31.5°C, so that heritability tended to increase. Richards (1978) examined grain yield in two rapeseed species under drought and irrigated conditions using 28 half-sib and 112 full-sib families. Yield of *Brassica campestris* was decreased by 25 per cent under drought conditions. Broad heritability was similar in the two environments, but narrow heritability was higher in the unstressed environment. Yield of *B. napus* was decreased by 68 per cent, and both the narrow and broad heritabilities were higher under stressful conditions.

Changes in the genetic components of variance may not only be confined to the additive and dominance components, but may also involve epistasis. Jinks *et al.* (1973) examined genetic components for growth rate, height, flowering time, and leaf length in *Nicotiana rustica* from crosses in different environments and found that epistatic interactions tended to be important under unfavourable conditions but not in favourable environments. Unfortunately, the generality of this finding is not known because epistatic interactions cannot be detected in many of the experimental designs used to examine the effects of stress on genetic variation in plants.

How can these different results be reconciled? The limitations discussed above for agricultural animals apply to many of the plant experiments. A potential problem with comparisons over a range of localities is that stress conditions are not defined, and this will decrease the heritability estimates for performance under stressful conditions (Buddenhagen 1983). For example, plants growing in unfavourable localities may experience several different types of stress, and yield differences between cultivars will be reduced if one set of cultivars is more resistant to one stress while a second set of cultivars is more resistant to a different stress. This criticism does not apply to experiments where only one stress is varied under controlled conditions (e.g. Richards 1978), and it is worth noting that those studies reporting an increase in the genetic variance with stress involved defined changes in a single environmental variable. Another limitation is that some experimental designs (e.g. comparisons of varieties) do not allow dominance and additive components of the genetic variance to be distinguished. It is clear that these components can vary with stress levels, so that only tentative inferences about narrow heritability can be made from varietal comparisons.

Little is known about changes in genetic variance within natural plant populations that have not been exposed to artificial selection for traits of agronomic importance or laboratory selection. Some genecological studies indicate that ecotypes with the same morphology under field conditions may show morphological differences when they are grown in a common garden under favourable conditions (Bradshaw 1965), suggesting an increase in heritability for morphological traits when stress is reduced.

This section can be summarized by the factors, listed in Table 5.5, that contribute to the diversity of approaches used to consider changes in genetic variance components with stress levels in animals and plants. It is not surprising that results from different studies have often been inconsistent. Some of the inconsistencies arise from the broad definition of 'stress' in the various studies: small decreases in productivity or growth were considered stressful in some studies, while stressful conditions in other studies represented conditions near lethality. Most experiments examined only two levels of an environmental factor, which may be

Table 5.5. Factors contributing to the diversity of experimental designs that test the association between genetic parameters and stress levels.

---

1. Strain comparisons or comparisons of different generations or selection experiments.
2. Experimental subjects cultured under different levels of stress or cultured under non-stressful conditions.
3. Stressful conditions defined by levels of an environmental factor or by mean productivity.
4. Natural populations, laboratory strains or agricultural populations provide test organisms.
5. Variation in the severity of a stress.
6. Division of genetic variance into additive, dominance and epistasis components, or components not distinguished.
7. Types of traits examined, in particular yield versus survival.

---

inadequate to detect changes in variance components at environmental extremes. For example, Westerman and Parsons (1973) found that the additive genetic variance for longevity only increased markedly when an extreme level of radiation was used (Table 5.3). The distinction between experiments where organisms are cultured in favourable environments and where they are only being tested under stress also seems important. More experiments are required that consider defined levels of stressors important in natural populations that include conditions approaching lethality.

5.5 Changes in heritability with stress: explanations

What processes account for changes in genetic parameters with increased stress? The first of several possibilities (Table 5.6) is that decreases in heritability with increasing stress levels reflect an increase in the environmental variance because of the exaggerated effects of small environmental inputs on growth and yield under stressful conditions.

Table 5.6. Factors contributing to changes in heritability and genetic components of variance with stress levels.

---

1. Change in environmental variance with stress levels.
2. Past history of directional selection or stabilizing selection.
3. Changes in the association between metabolic flux and enzyme activity because of substrate availability or the number of genes affecting flux levels.
4. Decrease of canalization under stress.
5. Increase in recombination or mutation rates with stress.
6. Different genes act at different stress levels.

Blum (1988) argued that when plants are stressed, small increases in factors such as water and nutrients can result in large growth responses. Similar inputs will not have the same effect on growth under more optimal conditions. This means that the same level of environmental variation under stressful conditions will have a much larger influence on fitness traits, inflating the error variance. This explanation is relevant to cases where the environmental variance increased with stress, and may explain the decreased heritability under stress in some of the experiments on plants. However many of the heritability changes were associated with alterations in the genetic variance, which does not explain changes in the relative size of the dominance and additive genetic components.

Another possibility is that heritability differences reflect the past history of selection on a trait. If a trait is under directional selection, then favoured alleles are likely to become fixed in a population, decreasing the additive genetic variance. The intensity of directional selection will depend on environmental conditions. The additive genetic variance of a fitness-related trait might be expected to increase as stress levels approach conditions that are rarely experienced in nature, because genes that contribute to the variance only at extreme levels may be under intermittent directional selection. In contrast, less extreme climatic stresses will often occur because of seasonal variation. The additive genetic variance should therefore be lower under conditions that are commonly encountered in nature, and increase under stress levels that are rarely encountered. This would help to explain why the additive genetic variance often increases at extreme levels of stress in many animals. However this argument does not apply to stresses that are not commonly encountered in nature, such as radiation and phenylthiocarbamide. In addition, some of the 'optimal' environments used in laboratory experiments may rarely be encountered in the wild.

If selection reduces the additive genetic variance, this should be evident in comparisons of populations exposed to different levels of environmental stress. The results in Table 5.7 suggest that populations of *D. melanogaster* from Melbourne and Darwin, with high resistance to desiccation and ethanol, and those from Townsville, with a much lower resistance (Section 4.2), differed in their level of genetic variation, reflected in the variance among strains and coefficients of variation. The lower genetic variation in Melbourne and Darwin is in accordance with a likely history of directional selection in these populations because of harsher climatic conditions (Parsons 1980).

Mather (1943, 1966, 1973) argued that selection histories can also influence the size of the dominance and additive components of variance, based on Fisher's (1930) theory of dominance. Fisher noted that mutant

Table 5.7. Variance among strains (VS) and coefficients of variation (CV) for the adult desiccation resistance and ethanol resistance of *D. melanogaster* isofemale strains from three Australian populations. (Modified from Stanley and Parsons 1981.)

		Melbourne	Townsville	Darwin
Desiccation resistance				
Males	VS	2.098	3.340	1.032
	CV	7.8%	11.9%	5.7%
Females	VS	1.134	3.415	1.438
	CV	4.6%	9.3%	5.4%
Ethanol resistance				
	VS	0.0147	0.0178	0.0121
	CV	11%	78%	20%

Values are based on the analysis of 8 Melbourne strains, 9 Townsville strains and 10 Darwin strains.

alleles are usually recessive and are therefore not expressed in the heterozygous form. Such alleles would only tend to increase slowly in the population if they were favoured by selection, but would increase more rapidly if they became associated with dominance modifiers. Favoured mutant alleles were therefore expected to become dominant as a consequence of directional selection. Mather extended this idea to quantitative traits, by suggesting that these traits are likely to exhibit dominance in the direction of the favoured alleles affecting them, and exhibit additive genetic effects when traits are not under directional selection. Epistasis is also expected to be more important under conditions of directional selection, because two non-allelic genes that jointly have a greater influence than the sum of their individual effects should be favoured under directional selection when they increase the expression of a trait in the favoured direction.

Such arguments have been used to explain changes in genetic variance with stress levels. If a trait is under directional selection, then directional dominance and epistasis should be important, whereas the additive genetic variance should become more important when a trait is not strongly selected or when it is under stabilizing selection. Jinks *et al.* (1973) have interpreted their data on *Nicotiana* from this perspective, suggesting that traits that are normally under stabilizing selection may be under directional selection in extreme environments where their expression will be altered the most. For example, consider a trait such as flowering time which is under stabilizing selection because plants flower at a particular time of the year to maximize exposure to suitable pollinators and seed dispersal agents. If environmental conditions delay the flowering period, selection will become directional for early flowering while the reverse will occur if conditions cause early flowering. This leads to the expectation of epistasis or directional dominance under stressful conditions, but not under favourable conditions. Although Jinks *et al.* (1973) found evidence for epistasis under stress they did not provide evidence that intermediate phenotypes of the traits showing epistasis under stress are selected.

A problem with this approach is that Mather's dominance arguments may only apply to some alleles. Wright (1934) noted that dominance of many alleles might be expected on biochemical grounds without the need to invoke directional selection, and Charlesworth (1979) found that the degree of dominance of lethal genes was not consistent with predictions from Fisher's theory. It seems likely that dominance will often follow directly from the association between flux and enzyme activity, as discussed below. Thus while there is some empirical support for an association between directional dominance and the type of selection likely to be acting on a trait (Broadhurst 1979), this association may not necessarily reflect a history of directional or stabilising selection.

A third explanation for the changes in genetic variance components with stress is based on the Kacser and Burns (1981) model of the relationship between enzyme activity and flux, outlined in Fig. 5.6 and Section 5.3. Flux is more closely related to fitness than enzyme activity. Dominance for fitness traits may result from this relationship because flux associated with a heterozygote with intermediate enzyme activity will be similar to the flux for a homozygote with the highest enzyme activity. Changes in catalytic activity at one step of a metabolic pathway tend to be buffered by other steps: the larger the number of enzymes in a pathway, the greater the buffering, and the greater the degree of dominance for an allele with high activity. Alleles that influence enzyme activity may therefore only have additive effects on the phenotype (flux) if they are involved in short biochemical pathways. These considerations suggest that the recessive nature of mutants may be a consequence of the interaction of biochemical pathways as well as the evolution of dominance modifiers. Expressed in this way, the theory of dominance outlined by Mather may not be required to explain directional dominance for fitness components. Mutant alleles are likely to be recessive, regardless of whether or not wild-type alleles have been subjected to directional selection.

This model can account for changes in the genetic variance with stress levels if it is assumed that the nature of the association between flux and enzyme activity is altered. We have already discussed the possible increase in the sensitivity coefficient under stress so that this association becomes more linear and differences between genotypes become evident (Section 5.3). A more linear relationship will increase the additive genetic component. In addition, the degree of dominance for increased flux will change, as discussed with respect to heterosis. Such changes could occur under stress if enzyme substrates become limiting or if the number of loci affecting flux decreases. Unfortunately, we know little about the effects of stress on metabolic processes, although changes in adenylate energy charge indicate that metabolic alterations take place. Some enzymes associated with stress resistance may show a linear relationship between flux and activity. For example, alcohol dehydrogenase is correlated with ethanol resistance in *Drosophila melanogaster*, and the sensitivity coefficient which defines the change in flux with enzyme activity may be close to 1.0 for this enzyme with respect to the conversion of ethanol into lipid (Geer and Heinstra 1990).

The association between stress, canalization, and asymmetry provides a fourth explanation for changes in the dominance variance and additive variance components with stress levels, at least in experiments where organisms are cultured under stressful conditions. These conditions may release phenotypic variability by decreasing the stability of the developmental system. This could increase the genetic variance if there is a genetic component to the increased phenotypic variability. The fact that phenotypes such as crossveinless and bithorax could be selected by Waddington and others after stress exposure indicates that the production of these phenotypes has a partial genetic component. Arking and Clare (1986) suggested that canalization reflects the buffering evident in the relationship between metabolic flux and enzyme activity, discussed above. Changes in flux are essentially buffered from changes at enzyme loci, so canalization may be considered an inherent property of *in vivo* systems that are not perturbed by stress. Arking and Clare argued that this explanation could account for the successful selection of longevity in *D. melanogaster* under crowded conditions.

The increase in recombination under stressful conditions, discussed in Section 5.1, probably has little effect on changes in the genetic variance with stress levels. Considering two loci, Mather and Jinks (1971) derived equations for the variance in the F_2 under variable levels of recombination when alleles at linked loci increasing the expression of a trait are in a coupling arrangement (both alleles on same chromosome) or a repulsion arrangement (alleles on different chromosomes), and argued that genetic variance will be increased when genes are in a repulsion arrangement and decreased in a coupling arrangement, because recombination can produce extreme phenotypes in the former situation. However this was not confirmed in a more general analysis by Turelli and Barton (1990) which made no assumptions about the number of loci or alleles influencing a quantitative trait. These authors found that recombination had little effect on the equilibrium genetic variance of quantitative traits that are under stabilizing selection. Increased recombination as a consequence of stress may be important when genetic components are characterized from directional selection experiments. Directional selection causes alleles at linked loci to form a repulsion arrangement and recombination can increase the genetic variance under these conditions (Maynard Smith 1988), although this has not been verified experimentally.

Mutation may have some impact on genetic variability in long term selection experiments. 'Mutations' associated with unequal crossing over in the rDNA region of *Drosophila melanogaster* can be an important source of variation in selection experiments for quantitative characters (Frankham 1980, 1988) and increased mutation rates under stress could therefore increase realized heritabilities. Hill (1982) has demonstrated theoretically that a long term continuous response to artificial selection may be accounted for by new mutations, but there are no empirical data for stress response traits.

Most of these explanations predict an increase in the additive genetic variance with increasing stress levels. The genetic variance may also increase or decrease with increasing stress because different genes contribute to the same trait measured in different environments. There is evidence that the same genes do not influence variation in a trait across all stress levels (e.g. Yamada and Bell 1969; Richards 1978; Rumbaugh *et al.* 1984), as will be considered further in Section 7.3. Genes contributing to a trait in one environment do not necessarily contribute identically to variation in the same trait in another environment, as demonstrated for genes affecting bristle number in *D. melanogaster* (Schnee and Thompson 1984) and the response of plant genotypes to different SO_2 levels (Hutchinson 1984). Genes that do not contribute to resistance under low stress levels may be largely responsible for resistance at higher stress levels, and this could increase or decrease the genetic variance.

In summary, several factors can account for changes in genetic variance components with stress levels (Table 5.6). The relevance of these factors to particular situations depends to some extent on the experimental methods that have been used to examine these changes (see Table 5.5), as well as on the history of selection on a trait.

5.6 The number of loci affecting stress responses

How many loci affect natural variation in stress response traits? There are two aspects to this problem, the number of loci that actually contribute to genetic variation within a population, and the number of loci contributing to changes in the mean of a trait as a consequence of directional selection. These may not be equivalent, because some of the genes involved in a selection response may be rare in the base population, whereas genes at intermediate frequencies will be responsible for most of the variation within a population.

The number of loci influencing stress response traits will depend on the nature of the resistance or evasion mechanism. Resistance to specific chemical stresses such as insecticides may involve a simple genetic basis; it can be acquired by relatively simple mechanisms, such as structural changes in the target enzyme of a toxin, gene amplification or changes in a detoxification pathway. On the other hand, climatic stresses such as desiccation and temperature extremes may have multiple effects, and increased resistance could involve genetic changes at numerous loci. The actual number of loci involved will depend on the stage at which a mechanism acts. Mechanisms that keep a stress out may be relatively simple and be determined by a few genes. Once a stress enters an organism it often has multiple effects so that a large number of genetic changes may be required to achieve a high degree of resistance. Genes that are involved with 'environmental filtering' (Watt 1985) may help to exclude stress effects from an organism. As an example, salt resistance in plants may be due to a specific filtering mechanism such as the exclusion of salt, or accumulation of ions, mechanisms which may be determined by a single gene (e.g. Abel 1969). The demonstrable fitness differences for electrophoretic variants of enzymes involved with environmental filtering also indicate their large potential effect on the genetic variance (Section 3.3). Mechanisms involving changes in metabolic processes such as patterns of energy allocation will probably have a complicated genetic basis although large regulatory effects can be exerted by single loci for enzymes at some positions in metabolic pathways (Watt 1985).

The intensity of selection is another factor influencing the number of loci involved in a response to natural or artificial selection. If a large number of common genes with small effects are segregating in a population, as well as a few major genes with large phenotypic effects, then genetic changes via the major genes may be favoured under intense directional selection (Lande 1983). This factor probably accounts for the difference in the number of genes controlling insecticide resistance in the field and in lines selected under laboratory conditions (Roush and McKenzie 1987). Field populations tend to show monogenic resistance, whereas the response to selection in laboratory experiments tends to have a polygenic basis. The experiments of Crow (1957) and King and Somme (1958) on DDT resistance in *D. melanogaster* are often given as examples of a polygenic response, because factors on all chromosomes contributed to the increased resistance of a selected line. Most of the laboratory selection regimes have used mortality levels of 80 to 90 per cent. These levels are likely to select within the phenotypic distribution of the base population determined mostly by polygenes, because alleles with major effects on insecticide resistance tend to have deleterious effects (Section 7.2) and are, therefore, likely to be rare in unselected populations. In addition, laboratory selection experiments tend to involve small populations, making the occurrence of rare major genes unlikely. In contrast, field populations of insects are much larger, making the occurrence of rare alleles with major phenotypic effects on resistance likely. In addition, the dose of insecticide applied to a field population will be high, because mortality levels approaching 100 per cent are required for pest control, and individuals with major genes at the extreme of the phenotypic distribution will be selected. This is also an example where genetic variation contributing most to the genetic variance in the base population is unimportant in genetic changes under directional selection. These considerations may apply to other stress response traits under directional selection where selection intensities are high, although many stressful situations which are considered extreme involve mortality levels that are similar to those used in artificial selection experiments (Chapter 2).

Not much is known about the genetic basis of stress resistance traits other than insecticide resistance. Large differences in resistance have been associated with one or a few genes or have been localized to a chromosome. We have already considered some examples where large differences in stress resistance were associated with variation at single enzyme loci (Chapter 3). In addition, studies that have attempted to dissect resistance differences between strains have often uncovered a simple genetic basis. Qualitative differences in heat resistance between two strains of the nematode *Caenorhabditis elegans* from England and France segregate as a single locus (Fatt and Dougherty 1963). The French strain is only able to mature and reproduce normally up to 18°C, whereas the English strain grows normally in the range 10 to 25°C. In *Drosophila*, there have been some attempts to localize variation in resistance to the chromosome level. Ogaki and Nakashima-Tanaka (1966) found that the difference in radiation resistance between a wild strain and a mutant strain of *D. melanogaster* involved one or more genes on chromosome 3. Parsons *et al.* (1969) localized differences between radiation-resistant and sensitive strains to various parts of chromosomes 2 and 3. Ogaki *et al.* (1967) developed an ether-resistant strain in which the third chromosome was mainly responsible for the resistance, with major gene activity at position 61, and minor activity on the X and fourth chromosomes. Morrison and Milkman (1978) obtained a rapid response to selection for heat sensitivity in an isofemale strain of *D. melanogaster* where almost all of the selection response involved the second chromosome. As a final example, Duke and Glassman (1968) attributed differences in resistance to streptomycin and fluorouracil between *D. melanogaster* strains to the third chromosome. Unfortunately, locating genes to an entire chromosome in *D. melanogaster* does not provide much information on the number of loci controlling a trait because of the low chromosome number in this species. Several genes could be involved if variation maps to a single chromosome, while only four genes may be involved if each chromosome contributes to a selection response.

Crosses between susceptible and resistant varieties of crop plants often suggest a simple genetic basis. The review in Blum (1988) provides examples of resistance to drought, chilling, and mineral deficiencies. There are also cases where segregation data suggest the involvement of more than one gene, although it becomes difficult to distinguish between varietal differences influenced by two or three genes or numerous genes on the basis of crosses because both explanations often provide an adequate fit for segregation data. Moreover, segregation data only provide an estimate of the 'effective' number of loci; the actual number may often be much larger (Lande 1981).

Segregation ratios suggest that heavy metal resistance in plants is probably controlled by a few genes (Macnair 1981). The genetic basis of copper resistance in the yellow monkey flower, *Mimulus guttatus*, was studied by crossing an individual from a resistant population with one from a non-resistant population (MacNair 1983). Segregation ratios provided evidence for a single major gene. This study indicates that a single gene can account for resistance differences between populations, but does not prove that only one resistance gene is involved because crosses were not carried out with different individuals from the same populations. It is possible that individuals carrying resistance alleles from different loci show the same level of resistance. For example, different mutants in *Caenorhabditis elegans* affect sensitivity to the anaesthetic halothane, but produce the same phenotype (Sedensky and Meneely 1987).

These findings suggest that large differences in resistance can be attributed to single chromosomes and perhaps to single genes, but do not prove that variation in natural populations or responses to directional selection are largely due to variation at a few loci. To demonstrate this, gene location studies and segregation analyses would need to be carried out on a random sample of strains from a population instead of two or more strains with extreme phenotypes. This would rule out the possibility that genetic variation was due to a number of major genes at a low frequency in the population (Hoffmann and Parsons 1988). Unfortunately, extensive studies on the identity of genes segregating in different strains have only been carried out for a few quantitative traits, such as Milkman's work on crossveinless in *D. melanogaster* (Milkman 1965, 1979).

Turning to stress evasion traits, most studies on variation within populations have not extended beyond the estimation of heritability (Section 4.5). In insects, crosses between populations or sibling species for diapause characters indicate Mendelian segregation in some cases but the absence of clearly defined segregation patterns in other cases (reviewed in Tauber *et al.* 1986). Variation in the production of winged morphs in water striders was considered due to a single locus (Vepsalianen 1978), although the evidence for this was not conclusive (Harrison 1980).

A general problem with many genetic analyses of selected lines or populations differing for quantitative traits is that they are usually unable to distinguish between a few loci and a large number of loci (Barton and Turelli 1990). Moreover, most studies have been restricted to morphological traits correlated with body size, or fitness measurements under optimal conditions that represent the cumulative effects of numerous traits, and are therefore likely to have a complex genetic basis. We can therefore only say that single genes may have large effects on resistance, and that changes in the mean of the population may occur via changes at a few loci or many loci. This unsatisfactory situation applies to quantitative traits in general (Barton and Turelli 1990).

5.7 Persistence of genetic variation

Accounting for levels of genetic variation in natural populations has been a central preoccupation of evolutionary geneticists for many years (Lewontin 1974). This remains true although the emphasis is shifting from a preoccupation with the maintenance of genetic variation detected by electrophoresis to discussions on levels of quantitative genetic variation (Barton and Turelli 1990). Some of the models for maintaining genetic variation are relevant to genetic variation in stress responses, and many invoke the types of fitness costs that will be discussed in Chapter 7.

5.7.1 Heterozygote advantage

Genetic variation will persist at a locus if there is some form of heterozygote advantage, for example if heterozygotes have greater stress resistance than homozygotes in an organism continuously under stress. There are few examples of heterozygote advantage for increased stress resistance in the literature. There is evidence for an association between heterozygosity and starvation resistance in *Mytilus edulis* (Diehl *et al.*

1986), but it is not clear if this reflects heterozygote advantage at individual loci or other explanations for heterosis, discussed earlier in this chapter. Many experiments indicate that genetic variance for stress resistance within populations is largely additive or shows partial dominance rather than overdominance (Section 5.4), which suggests that the potential for heterozygous advantage contributing to differences between genotypes is small. There are cases where stress resistance is controlled by a single locus and where heterozygote advantage is absent, such as in alleles responsible for insecticide resistance. An exception is variation at the *Lap* locus in *Mytilus edulis*, where heterozygotes for one of the alleles have higher enzyme activity, associated with greater amine uptake and excretion rates under low salinity conditions (Hilbish and Koehn 1985). In addition, Nevo (1988) cited two allozyme loci in marine organisms where heterozygotes had higher fitness at intermediate levels of pollutants, although the cause of the fitness differences was not known.

5.7.2 Genotype–environment interactions and induced overdominance

Heterozygote advantage can also arise indirectly because of genotype–environment interactions where the relative fitness of different genotypes depends on the environment. Single locus models associated with environmental variation in time (temporal) and space were reviewed by Felsenstein (1976) and Hedrick *et al.* (1976). When the environment varies in time and the relative fitness of the homozygotes in different environments switches, polymorphism in a diploid population can be maintained if the heterozygote has the highest geometric mean fitness (Gillespie 1973). The geometric mean is sensitive to variability in fitness values, and heterozygote advantage arises from the fact that the fitness of the heterozygote is less variable than that of the homozygotes. Polymorphism can be maintained even when the fitness of the heterozygote is intermediate to that of the homozygotes in all environments. Hence temporal environmental variation can maintain genetic variation as long as different homozygotes are favoured in different generations. Variation will not be maintained if one type of environment persists for several generations, so that one allele becomes fixed. Environmental heterogeneity must therefore be common or must occur cyclically for the persistence of genetic variation. The conditions for maintaining genetic variation by this mechanism become more restricted in an age-structured population (Orzack 1985), and will depend on the frequency of the different environments and the autocorrelation between environments.

Reviews of single locus models maintaining variation by spatial heterogeneity are given in Hedrick *et al.* (1976) and in Hedrick (1986). Most models can maintain genetic variation only under restrictive conditions. In particular, the sizes of the environmental patches have to be similar. If it is assumed that one homozygote is favoured in one patch, and another homozygote is favoured in a different patch, then patch sizes have to be extremely similar unless selection is strong or unless genotypes 'select' habitats wherein they have highest relative fitness. The only exception is the 'SAS–CFF' model, described in Gillespie (1978), in which the fitness of the heterozygote in a patch is close to that of the fittest homozygote. This means that the arithmetic mean fitness of the homozygotes over all patches is lower than that of the heterozygote.

More recently, attention has focused on quantitative genetic models of environmental heterogeneity to see if these can maintain substantial heritabilities. Via and Lande (1987) approached this question using genetic correlations to examine the performance of genotypes in different environments. This approach will be discussed further in Section 7.3. They concluded that genotype–environment interactions would not maintain genetic variance unless the genetic correlations were close to + 1 or − 1, and showed that a single optimum genotype would generally be favoured. As pointed out by Gillespie and Turelli (1989), this result is not particularly surprising because a genetic correlation of less than unity means that increased performance in one environment is only partly offset by decreased performance in another environment, so that adaptation to the two environments can proceed partly independently. Eventually, a single genotype could theoretically evolve that is superior to all others in all of the environments. In contrast, Gillespie and Turelli (1989) started with the assumption that a single genotype cannot perform the best across all environments, even though the same phenotype is favoured in a trait under stabilizing selection in all environments. If alleles have different additive effects on the trait in different environments, the heterozygotes have a lower phenotypic variance across environments and therefore have higher mean overall fitness. This will maintain substantial levels of additive genetic variance, regardless of whether the environmental heterogeneity is spatial or temporal.

Does environmental heterogeneity contribute to genetic variation in stress response traits? Opposing selection, where one genotype is favoured in one environment and a different one is favoured in another environment, has often been invoked to account for the maintenance of variation in stress evasion traits such as dormancy, although interpretations have often been made from the perspective of group selection rather than individual selection. A low number of dormant individuals in a population is often considered to be a mechanism that enables a population to survive periods of stress severe enough to kill the non-dormant individuals (e.g. Takahashi 1977; Tauber *et al.* 1986). Genetic variation in dormancy may also be accounted for by models of individual selection incorporating environmental heterogeneity. Dormant genotypes may have higher fitness in one set of conditions and non-dormant genotypes may be favoured under other conditions. Phenotypes that enter a prolonged diapause or some other dormant state will be favoured under the stressful conditions which occur occasionally because non-dormant phenotypes may not survive. However non-dormant variants will normally be favoured in the absence of adverse conditions because they continue to grow and reproduce. In addition, genotypes producing dormant stages may be at a selective disadvantage under favourable conditions because there are energetic costs (Section 7.1). Similar arguments can be made for other stress evasion traits, such as migration from unfavourable conditions. This means that variation in these types of traits may be maintained by models of environmental heterogeneity because the relative fitness of genotypes can switch.

Single locus models of temporal variation may be relevant to some insect diapause characters which show simple patterns of Mendelian segregation (see examples given in Tauber *et al.* 1986). Unfortunately, assigning fitness values to genetic variants is difficult because their success needs to be considered over a period of several years. Quantitative genetic models are more appropriate for life-history traits involved in stress evasion. Newman's (1988*a,b*) study of development time in toads (Section 4.5) provides a good example, where alleles increasing development time were favoured under one set of conditions (ponds dry up slowly) but selected against under different conditions (ponds dry up rapidly). Grant and Grant's (1989) study of Darwin's finches in the Galapagos provides another example where different morphological phenotypes were selected under a variable food supply. Long-billed birds were favoured when the food sources that Darwin's finches normally exploit (*Opuntia* flowers and fruit) were absent, while birds with deep beaks were favoured in another period when the predominant food available consisted of arthropods beneath the bark of trees and from *Opuntia*.

Models of spatial and temporal variation are relevant to genetic variation in stress resistance traits as well as evasion traits. There is some evidence that the same genotypes are not favoured in environments that encompass both optimal and stressful conditions (Chapter 7). Genotypes with high levels of stress resistance may tend to perform poorly under optimal conditions, so that extreme stress periods interspersed with favourable periods can be associated with changes in the fitness rankings of genotypes as envisaged in models of environmental heterogeneity.

It is not clear whether the requirements of the SAS–CFF model can be met by variable conditions of environmental stress. Some of the data discussed in Chapter 3 suggest that different enzyme morphs are favoured under different environmental conditions so that reversals of genotype fitness could occur, and heterozygote fitness should be closer to that of the fitter homozygote if the association between enzyme activity and flux follows Fig. 5.6 and flux is related to fitness. Dominance for stress resistance traits tends to be in the direction of increased resistance (Section 5.4), as required by this model. Models of geometric mean overdominance and the Gillespie and Turelli (1989) model for quantitative traits require that the performance of heterozygotes is less variable across environments. This is commonly found in studies on changes in heterosis with stress levels, where the fitness of hybrids declines relatively more slowly than that of the parental strains as conditions become stressful.

The above discussion has considered spatial variation within a population, but genetic variation may also be influenced by spatial variation between populations that are linked by migration. Genetic variation can persist if one genotype is selected in one population and an alternative genotype is favoured in a different population, and there is a low rate of migration between the populations. Single-locus models are discussed in Felsenstein (1976). Such models may be relevant to the examples of spatial variation for stress resistance discussed in Section 4.4, which indicate that genetic differences between adjacent habitats can arise even when there is gene flow between the habitats. Gene flow ensures that some genetic variation for stress resistance persists in both habitats.

5.7.3 Pleiotropy and induced overdominance

Genetic variation may be maintained if genes have pleiotropic effects on more than one fitness-related character and alleles associated with high fitness for one character have low fitness for another character (Rose 1982). This process of 'antagonistic pleiotropy' can maintain variation when the alleles that are associated with increased fitness for at least one of the traits are dominant, because this will result in overall overdominance.

Findings which will be considered in Chapter 7 suggest some cases where alleles increasing stress resistance or increasing the ability to avoid a stress are associated with decreased performance for another fitness-related trait. For overdominance to occur, at least one of these traits must show directional dominance for increased fitness. Some of the studies discussed above indicate that, when it occurs, dominance for resistance traits is in the direction of increased resistance, so dominance will be expressed under stressful conditions. In addition, alleles with low stress resistance but relatively high fitness under optimal conditions may be dominant in favourable environments because there is evidence that many fitness-related traits measured under optimal laboratory conditions show strong dominance in the direction of increased fitness (Mather 1973; Broadhurst 1979). Insecticide resistance provides a specific example: many resistance alleles tend to show deleterious fitness effects in the absence of the insecticide only when they are in the homozygous form (Roush and McKenzie 1987), indicating directional dominance for increased fitness. Conditions for the maintenance of genetic variation by antagonistic pleiotropy may therefore be applicable to some stress response traits.

Antagonistic pleiotropy can also occur between traits that are expressed at different life cycle stages. This type of selection may be relevant to situations where one stage is more susceptible to stress than another stage. For example, many plant seedlings are more susceptible to environmental stress than mature individuals, as reflected in the high mortalities experienced by most plants at the seedling stage (Cook 1979). Seeds that are stress resistant may be at an advantage over stress susceptible individuals in the establishment phase, but the relative fitnesses of these genotypes could switch at the adult stage because of the lower competitive success of resistant genotypes once they are no longer exposed to a stress (see Chapters 6 and 7). Juvenile stages of many animals are also relatively more susceptible to stress, as indicated by the high juvenlie mortalities of different species during the recent El Niño event (Section 2.2).

5.7.4 Mutation–selection balance

One explanation that is often invoked to account for high levels of quantitative genetic variation is mutation–selection balance. Levels of variation are considered to reflect a balance between new variation generated by mutation and variation removed by stabilizing selection. This idea was promoted by Lande (1975) as a general explanation for genetic variation in quantitative traits. The relevant theoretical models have been critically discussed by Turelli (1984, 1986). Problems arise in applying the theory because it is difficult to determine the intensity of selection and mutation rates for quantitative characters. Mutation–selection balance can only maintain substantial genetic variation if a trait is affected by a large number of loci or if the mutation rate for loci influencing a trait is high (Turelli 1984). As we do not have good estimates of the number of loci affecting stress resistance traits or mutation rates at these loci, it is difficult to evaluate the importance of this mechanism in generating variation. However the ability to locate genes for some stress resistance traits, and the simple genetic basis for some stress avoidance traits, suggest that the number of loci will vary from trait to trait and, in many cases, may not be in the range required for the maintenance of a large amount of variation. More estimates for these parameters are required, particularly under stressful conditions that generally enhance mutation rates (Section 5.1).

5.7.5 Concluding remarks

The above discussion suggests several explanations for the maintenance of genetic variation in stress response traits. Spatial and temporal variation may be particularly important because stress resistant genotypes are not favoured in all environments. Genetic variation in stress response could also be influenced by antagonistic pleiotropy, and could persist in some situations because of migration between environments that are associated with different levels of stress.

5.8 Summary

Recombination frequencies and mutation rates tend to increase under stressful conditions; rates of evolutionary change could therefore be enhanced in adverse environments. Developmental instability, in particular fluctuating asymmetry, may be enhanced under both environmental stress and genomic stress. This could increase variability under stressful conditions.

Heterosis tends to increase with stress, except in response to poor nutritional conditions. The increase may reflect an interaction between inbreeding and stress, heterozygote advantage or dominance effects averaged over a number of loci.

The heritability of many traits tends to increase with increasing stress levels in animals, although exceptions do occur. Heritability also changes with stress levels in plants although there are no consistent trends. The absence of any generalizations may reflect the diversity of experimental approaches, difficulties in defining actual stress levels, and the range of organisms (domestic and wild) that have been used to investigate the association.

The relative importance of dominance, additive, and epistatic components of the genetic variance often changes with stress levels. Additive effects tend to be higher at extreme stress levels. Resistance tends to be dominant over stress susceptibility.

Changes in genetic components of variance with stress levels may reflect the past history of selection, metabolic interactions, changes in canalization or the expression of different genes at different stress levels.

Resistance to some stresses such as pesticides may have a simple genetic basis in field populations, but little is known about the number of loci affecting resistance to other environmental stresses except that single genes with large phenotypic effects can often be isolated.

Genetic variation in stress response traits may persist in populations because of induced overdominance from spatial and temporal environmental heterogeneity and because of antagonistic pleiotropy. Mutation–selection balance and heterozygote advantage may be less important for these traits.

6. General stress resistance

Resistance to environmental stress may involve mechanisms that are highly specific for one type of stress, such as detoxification pathways restricted to a particular class of insecticide, or may cover different stresses, and have a common physiological/biochemical basis. We have already encountered a general resistance mechanism in the form of heat-shock proteins that are induced by a variety of agents and may provide protection against a range of stresses (Section 3.2). If such general mechanisms are common, then individuals may have high or low resistance levels for a range of stresses, and selection for increased resistance to one stress will increase resistance to other stresses.

In this chapter we consider the evidence for general resistance mechanisms and genetic variation for such mechanisms. We start from an energetic perspective which provides an underlying general framework for understanding physiological processes, and examine the possibility that resistance to a range of stresses is associated with resource conservation. This is followed by a brief discussion of other mechanisms that may provide resistance to multiple stresses. Most of the empirical data on variation in general stress responses relate to the interspecific level, although a few studies have considered intraspecific variation at the genetic level by examining genetic correlations among traits. These correlations provide an indication of whether or not selection for resistance to one environmental stress will change resistance to other environmental stresses, and can be used to evaluate the relative importance of general versus specific resistance mechanisms in evolutionary responses to stressful conditions.

A major difficulty in looking for general resistance mechanisms is that phenotypic changes in response to an environmental stress may represent the effects of a stress rather than a resistance mechanism. For example, the production of ethylene by plants under stress may be a response to tissue damage. Alternatively, ethylene may function as a plant hormone and regulate the synthesis of compounds to counter the effects of stress (Ayres 1984). Variation in stress resistance may be correlated with many traits, but a causal relationship between these traits and resistance can only be established by detailed experimental studies. Some generalized responses to stress may only represent stress 'indicators' such as those discussed briefly in Section 1.3.

6.1 Stress, energetics, and metabolic rate

Increased stress resistance can be attained by two types of energetic changes. First, an organism can alter its energy expenditure patterns to counter the injurious effects of a stress. This may involve the diversion of energy from normal metabolic processes to other processes that protect the cell from injury or that repair damage. The reduction in the production of cellular proteins during the cellular response to a heat shock and the simultaneous induction of heat-shock protein synthesis (Section 3.2) presumably represents such a diversion of energy.

Secondly, a reduced ability to obtain resources under adverse conditions can be countered by the conservation of metabolic energy and other resources. This resistance mechanism is typical of many short term reversible responses to stress. For example, plants reduce their stomatal apertures in response to water stress to retain turgidity in the face of decreased water uptake. They live off energy reserves during the stressful period because decreased gaseous exchange through the stomata reduces photosynthesis. If the water stress persists, many plants drop their leaves to conserve water and to decrease metabolic costs associated with respiration. In ectotherms, increased resistance to high temperatures as a result of acclimation is associated with a decrease in metabolic rate; this response conserves metabolic energy because high temperatures normally increase metabolic rate (Bullock 1955). Energy conservation is taken to an extreme when organisms enter dormant stages which have extremely low metabolic requirements. We have already discussed dormancy as a form of stress evasion (Section 4.5), but it could also be considered an extreme form of resistance, and may occur in response to any unfavourable conditions where resource acquisition becomes restricted, such as a reduction in a food supply, or extreme climatic conditions that restrict growth and reproduction

Rates of energy metabolism have been measured in numerous species, including many that live under demonstrably stressful conditions. Metabolic rate is usually measured in animals by oxygen consumption which reflects the rate of oxidative metabolic processes. Oxygen consumption is often measured in the absence of activity and this is known as the 'resting' or 'maintenance' metabolic rate. A related measure is the 'standard' metabolic rate (also known as the fasting metabolic rate), which is the steady rate to which oxygen consumption declines in the absence of food. This rate is also known as the basal metabolic rate, although the term 'basal' implies that metabolic rate has reached the lowest level to which it can fall, which is not true in many situations (Schmidt-Nielsen 1984). It is well known that oxygen consumption increases as body mass increases, and this relationship can be described by allometric equations. Metabolic rate is normally corrected for mass, and the corrected value is known as the 'specific' metabolic rate. In ectotherms, metabolic rate varies with temperature and many other factors, so the conditions under which metabolic rates are measured need to be accurately defined.

Mammals inhabiting extreme environments often have a lower specific metabolic rate than expected (McNab 1986). A reduced rate will decrease food requirements, increase water conservation, and increase heat resistance because of lowered heat production. In an early comparative study, the oxygen consumption of 10 populations of *Peromyscus* representing nine subspecies was examined by McNab and Morrison (1963). Measurements were made in daylight when activity was inhibited. Rates were lower for desert–mountain subspecies than for individuals from a mesic habitat, and a further reduction was seen in solely desert subspecies. Differences were also found in temperature regulation, desert species being the poorest regulators at low temperatures, and the best at high temperatures. In a more recent study, Hinds and MacMillen (1985) considered oxygen consumption in 13 heteromyid rodent species in the thermoneutrality temperature range, when oxygen consumption is constant and not increased because of temperature stress. Species occupying the most arid areas (desert scrub) had lower mass-specific metabolic rates than other species of this family, while highest rates were found for the tropical species. The allometric relationships had similar slopes to those found for other eutherians, but this relationship was depressed in the xeric and semi-xeric species. Xeric species also showed lower evaporative water loss. The reduced levels of metabolism and water loss were interpreted as being adaptive for energy and water conservation. As a final example, Lovegrove (1986) found the resting metabolic rate in the social subterranean mole rat, *Cryptomys damarensis*, to be 29 per cent lower than predicted by the relationship between body mass and oxygen consumption for subterranean rodents in general. *C. damarensis* lives in arid areas, and other subterranean rodent species from arid or semi-arid habitats also had lower rates than species from mesic habitats (Table 6.1). The reduction in metabolic rate was interpreted as an adaptation to lower food and water availability in arid areas.

These findings from comparisons between species from arid and more mesic areas can be extended to invertebrates. Lighton and Bartholomew (1988) studied metabolic rate in the desert harvester ant, *Pogonomyrmex rugosus*, which lives in the Mojave desert of southwestern United States, an arid region with wide fluctuations in temperature and food availability. Individuals had only 39 per cent of the metabolic rate predicted from the allometric relationship for this group of organisms. Juliano (1986) compared metabolic rate and starvation resistance in two

Table 6.1. The resting metabolic rate (RMR) of 14 species of subterranean rodents as a percentage of that predicted by the allometric
rodent curve and the allometric subterranean rodent curve. (Modified
from Lovegrove 1986.)

Species	Habitat	% (A)	% (B)
Pitymys pinetorum	Mesic	134	171
Thomomys talpoides	Mesic	115	144
Tachyoryctes splendens	Mesic	90	113
Spalax leucodon	Mesic	89	112
Spalacopus cyanus	Mesic	89	112
Gromys pinetis	Mesic	88	111
Bathyergus suillus	Mesic	83	102
Geomys bursarius	Mesic	81	101
Bathyergus janetta	Semi-arid	78	97
Thomomys umbrinus	Semi-arid	77	97
Heliophobius argentocinereus	Semi-arid	75	95
Cryptomys hottentotus	Semi-arid	62	78
Cryptomys damarensis	Arid	57	71
Heterocephalus glaber	Arid	43	55

A: % RMR predicted from rodent curve: RMR $= 4.98\ M^{-0.331}$.
B: % RMR predicted from subterranean rodent curve: RMR $= 3.79\ M^{-0.322}$.

carabid beetle species, *Brachinus mexicanus* and *B. lateralis*. The former
species is dominant at temporary ponds associated with periods of food
and water stress, while *B. lateralis* dominates at more permanent ponds
where these stresses are less common. *B. mexicanus* had a lower resting
metabolic rate and was more resistant to starvation than *B. lateralis*.
Species differences in desiccation resistance were also found, but these
appeared to be related to resistance of low body water content and the
amount of expendable water rather than metabolic rate. Transpiratory
water loss and metabolism of carabid and tenebrionid beetle species
from arid areas in Africa were studied by Zachariassen *et al.* (1987).
Those species from xeric habitats that had high rates of water loss had
relatively higher metabolic rates under conditions that minimized
activity (Fig. 6.1). This trend was apparent in both groups of beetles,
suggesting that the rate of oxygen uptake determines the rate of transpiratory water loss within this habitat group. The higher metabolic rate
of carabids compared to tenebrionids also correlated with their relatively higher rate of water loss, although other factors appeared to
contribute to the difference between beetles from dry and wet habitats.
    A reduced metabolic rate may also be characteristic of species living in
cold environments where food availability is restricted. For example, the

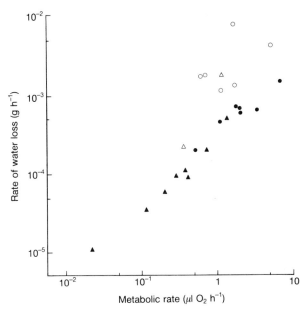

Fig. 6.1. Transpiratory water loss of carabids (circles) and tenebrionids (triangles) as a function of metabolic rate. Closed symbols represent beetles from xeric habitats, and open circles represent beetles from mesic or hygric habitats. (Simplified from Zachariassen *et al.* 1987.)

field metabolic rates of willow and crested tits, which overwinter at high latitudes in the northern hemisphere, were lower than predicted from the allometric relationship for these birds (Moreno *et al.* 1988). In contrast, coal tits had metabolic rates predicted by allometric equations; this species migrates to evade unfavourable periods, and has a more southerly distribution than the other species.

Low metabolic rates have been related to stress resistance in marine organisms. Coral species that were more resistant to heat in laboratory assays survived a temperature increase off the coast of Hawaii better than species that were less heat resistant (Jokiel and Coles 1974). The most sensitive of the seven species had the highest metabolic rate, while more resistant species had lower rates. Intertidal limpet species and other intertidal organisms that suffer periods of food shortage had low rates of oxygen consumption (Fig. 6.2), whereas those with abundant food had high rates of oxygen consumption (Newell and Branch 1980; Branch *et al.* 1988). The latter group could reduce their metabolic rate to some extent at times of food shortage, while limpets with low metabolic rates had a relatively constant rate even when food was provided.

These examples indicate that a low metabolic rate is a common feature

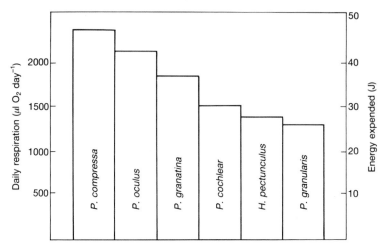

Fig. 6.2. Daily expenditure of metabolic energy of six limpet species. *Patella compressa* live on kelp and have abundant food; *P. oculus* and *P. granatina* live in the low to midshore at low densities and also have abundant food; *P. cochlear* occurs at high densities and is exposed to intense intraspecific competition; *P. granularis* and *Helcion pectunculus* occupy the high shore where food is limited. (After Branch *et al.* 1988.)

of species living in environments with decreased food availability and increased high temperature or desiccation stress. In many of these studies, species differences may be heritable because they persist when species are cultured under the same conditions. This association between metabolic rate and stress resistance may only be evident when resistance can be attained by energy conservation, and low metabolic rate will not provide resistance to some environmental stresses. In particular, organisms that are active in cold environments may have higher metabolic rates than those that live in warm environments when they are tested at the same temperature. Higher metabolic rates enable organisms to function at lower temperatures because the maintenance of electrochemical gradients and potentials across membranes is more expensive when it is cold (Hochachka 1988). Comparison of Antarctic and Californian mesopelagic fish (Torres and Somero 1988) showed that the Antarctic fish had a metabolic rate that was twice that of Californian fish when normalized to their temperature. This difference was found at the tissue level as well as the whole organism level, showing that activity levels of the organism were not involved. In contrast, high metabolic rates would not be expected in animals that hibernate to conserve energy in response to cold periods. Supporting evidence in lizards has been compiled by Tsuji (1988). Winter dormant lizards tend to reduce their metabolic rates after cold acclimation, whereas winter active species tend to increase it. This indicates species differences in the control of metabolic responses to environmental change, reflecting differences in phenotypic plasticity (Section 4.6).

6.2 Metabolic rate: intraspecific comparisons

Although metabolic rate has usually been considered at the interspecific level, there are some studies that address the association between stress resistance and metabolic rate at the intraspecific level. Two mammalian studies indicate that metabolic rate decreases with increased resistance to environmental stress. The resting metabolic rate of four karyotypes of the mole rat, *Spalax ehrenbergi*, varies with climatic conditions (Nevo and Shkolnik 1974), decreasing towards the desert, where conditions are relatively stressful in terms of heat and desiccation. Frisch (1981) selected a line of cattle to favour improved growth rate in tropical conditions. Normally, the growth rate was supressed due to parasitic infections, high temperatures, poor nutrition, and eye diseases. Bulls were selected on the basis of high post-weaning gain and control lines were maintained by random mating. Selection was carried out for a mean of 2.3 generations and performance results at two stress levels, one corresponding to cattle held in pens with low heat stress and food *ad libitum*, and the other corresponding to cattle in the field under stressful tropical conditions, are summarized in Table 6.2. Individuals from the selected line were more resistant to high temperature stress (as

Table 6.2. Selection responses in beef cattle selected for increased stress resistance, measured at two stress levels. (Modified from Frisch 1981.)

Variable	Stress level	Selected line	Control line
Gain/day (kg)*†	Low	0.79	0.85
	High	0.25	0.15
BIK score*‡	Low	3.15	3.82
	High	4.15	5.80
Rectal temperature*†§	Low	39.83	40.03
(°C)	High	39.47	39.98
Fasting metabolic rate*			
(kJ/kg/day)	—	83.3	89.1

*Significant differences between the means at each level of stress.
†Significant line by stress level interaction.
‡Level of infection by bovine infectious keratoconjunctivitis, an eye disease.
§Measure of heat resistance (lower rectal temperatures indicate higher resistance).

measured by rectal temperature) and parasites than control individuals. The selected line also had a lower mass-specific metabolic rate under fasting conditions, which was responsible for the increase in stress resistance since a reduced metabolic rate means less heat production and lower maintenance requirements. An increased resistance to parasites may arise because cattle are less resistant when they are at maintenance or losing weight than when they are growing, and growth can be sustained over a wider range of conditions when maintenance requirements are reduced.

Metabolic rate has also been associated with stress resistance in ectotherms. In the mussel, *Mytilus edulis*, Diehl *et al.* (1986) measured dry weight loss and oxygen consumption during starvation. Individuals were divided into large and small sizes on the basis of shell length, and both groups showed a positive correlation between the rate of standard oxygen consumption and dry weight loss, although the slope was steeper for the small mussels than for the large mussels (Fig. 6.3). Heterozygosity levels in the large group were significantly higher than in the small group, and there was an association between oxygen consumption

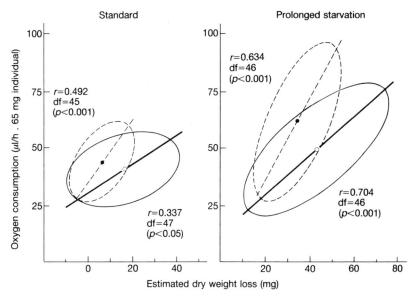

Fig. 6.3. Association between oxygen consumption and dry weight loss during starvation in *Mytilus edulis*. Mussels were divided into small (dashed line) and large (solid line) groups. One set of both groups was starved for 17–19 days before measurement (standard), while another set was starved for 53–63 days (prolonged starvation). Bivariate means are given, as well as the 95% confidence region as represented by the ellipses. (Adapted from Diehl *et al.* 1986.)

and heterozygosity within both groups, suggesting a genetic basis for the correlation with stress resistance (see Section 7.5). Tsuji (1988) considered the metabolic rate of the lizard, *Sceloporus occidentalis*, from two sites: southern California, where this species is active all year round, and Washington, where it hibernates in winter. Compared to the Californian population, the Washington lizards showed a depressed standard metabolic rate in winter and higher mass-specific standard metabolic rate in all other seasons, while those from California had an intermediate standard metabolic rate in winter. Acclimatization responses for metabolic rate therefore differed between these populations, with a lower metabolic rate during the stressful winter conditions experienced in Washington, although a genetic basis was not established.

There is evidence that metabolic rate is associated with genetic variation for stress resistance in *Drosophila melanogaster*. Eighteen isofemale strains showed a negative correlation between resistance to CO_2 exposure and standard metabolic rate (Matheson and Parsons 1973). This association was confirmed by a positive correlation between resistance to CO_2 and anoxia. The phenomenon may have been mediated by body size variation, since size was negatively correlated with metabolic rate and positively correlated with CO_2 resistance. Sierra *et al.* (1989) found an increase in CO_2 resistance and body size in *D. melanogaster* lines selected for resistance to acrolein, a common atmospheric pollutant. These correlated responses suggest that the selection response involved a reduction in metabolic rate: this was also indicated by a decrease in the locomotor activity of flies from the selected lines, since, in animals, activity levels are closely correlated with metabolic rate. Sierra *et al.* (1989) also found evidence for a non-specific resistance mechanism involving two enzymes that may be important in detoxification. We have carried out selection for increased desiccation resistance in replicate lines of *D. melanogaster* (Hoffmann and Parsons 1989a,b) and found decreases in the standard metabolic rate and activity levels of all three lines compared to unselected control lines (Table 6.3). The selection response was not associated with changes in body size. Finally, *D. melanogaster* lines selected for postponed senescence showed increased resistance to starvation, desiccation, and a toxic concentration of ethanol, and also had a decreased metabolic rate (Service *et al.* 1985; Service 1987). At 20 days old, however, flies from these lines differed in stress resistance, but not in metabolic rate, indicating that factors other than metabolic rate were involved in the altered stress resistance of old flies.

These results suggest that the interspecific findings on metabolic rate and stress resistance may occur at the intraspecific level, although more genetic studies are required on metabolic rate or other physiological

Table 6.3. Effect of selection for desiccation resistance in *Drosophila melanogaster* on desiccation resistance and metabolic rate: means and standard deviations for the selected and control lines. (Modified from Hoffmann and Parsons 1989*a*.)

		Time for 50% of the flies to die at 0% RH (h)	Metabolic rate (O_2/mg/h)
Selected lines	1	27.0 (1.4)	2.08 (0.15)
	2	28.4 (2.2)	2.15 (0.31)
	3	30.2 (2.6)	2.02 (0.32)
Control lines	1	17.8 (3.7)	2.74 (0.39)
	2	20.1 (4.1)	2.45 (0.52)
	3	17.8 (2.7)	2.65 (0.36)

variables likely to reflect energy conservation. There is less information on the biochemical basis of genetic changes in metabolic rate. Factors controlling the rate of respiration are fairly well known, at least in mammalian systems (Brand and Murphy 1987). Metabolic rate may be controlled at the gross level via changes in the number of mitochondria, or by more specific changes involving enzymes or cellular components affecting the overall rate of flux. Much of the control of flux within the mitochondria is exerted by the movement of protons, the relative concentrations of ATP and ADP, and the enzyme cytochrome oxidase. In turn, these may be affected by external factors in the cell, including the redox state of various energy carriers, which are influenced by the rates of reactions feeding electrons into the electron transport chain, and by the rates of reactions producing or consuming ATP. Control mechanisms can also be influenced by hormones which alter enzyme concentrations or enzyme kinetics. These controls suggest biochemical variables that might be associated with changes in metabolic rate, but they have not yet been included in studies of stress resistance and metabolic rate. Adenylate concentrations may be correlated with variation in the metabolic rates of tissues (Beis and Newholme 1975) and with acclimation responses (Walsh and Somero, 1981). Inter-individual and inter-specific variation in some enzymes, haemoglobin concentration and cardiovascular characteristics have been associated with metabolic rate in lizards, bullfrogs and anurans (see Pough 1989). These observations provide a starting point for biochemical genetic studies.

6.3 Slow growth and stress resistance in plants

Plant species exhibit varying rates of photosynthesis and respiration.

Genetically based differences among species or populations of the same species exposed to different ecological conditions can often be interpreted in an adaptive manner (Berry and Bjorkman 1980). For example, populations from cold climates tend to have higher rates of photosynthesis and respiration at low temperatures than populations from warmer locations (Mooney and Billings 1961; Bjorkman 1966). This leads to a minimization of the variation in growth rate among plants from a range of climatic conditions.

There are, however, inherent differences in growth rate between species from the same general area, and species with inherently slow growth rates tend to be found in more stressful habitats (Grime 1977, 1979; Chapin 1980). Slow growth rate was emphasized as a possible stress resistance mechanism by Grime (1979) in his classification of vegetation types (mentioned earlier; Section 2.4). He proposed three main types of plant 'strategies' on the basis of a review of the structure and composition of vegetation:

1. Competitors, which exploit conditions of low stress and low disturbance;
2. Ruderals, which are found under low-stress conditions in environments that are disturbed because of factors such as herbivores, pathogens, human activities, erosion and fire;
3. Stress tolerators, which are found under conditions of high stress and low disturbance.

Grime found that while stress tolerators are morphologically and taxonomically diverse, they show a conformity of life-history and physiology. These plants have inherently slow rates of growth, and are characterized by the evergreen habit, long-lived organs, sequestration and slow turnover of carbon, mineral nutrients and water, infrequent flowering, and mechanisms which allow the intake of resources during temporarily favourable conditions. Such characteristics may help in the conservation of captured resources in stressed environments where resources are scarce. Grime (1979) points out the existence of analogous strategies in stress-resistant fungi, which are characterized by slow-growing, relatively persistent mycelia and low reproductive effort. Examples include the slow-growing basidiomycetes, which form the terminal stages of fungal succession on decaying matter, and the lichens and ectotrophic mycorrhizas. Lichens in particular can grow under extremely unfavourable conditions.

A relationship between growth rate and adverse conditions can be demonstrated by comparison of similar species from different habitats within the same area. For example, Chapin et al. (1982) compared the growth rates of seedlings from two species of tussock grass at different phosphate concentrations. *Chionochloa crassiuscula* is found in soils with low phosphate levels, and *C. pallens* is found in soils with relatively high phosphate levels. *C. crassiuscula* grew more slowly at all rates of phosphate supply than *C. pallens*, and was less responsive to increased phosphate, as indicated by changes in weight and shoot height (Fig. 6.4). The two species had the same phosphate absorption capacity, which led to higher tissue phosphorus concentrations in *C. crassiuscula*. As another example, Grulke and Bliss (1988) compared the life-history characteristics of two grasses in the high arctic environment. *Phippsia algida* predominates in mesic sites and rapidly colonizes disturbed areas, while *Puccinellia vaginata* occurs in relatively undisturbed sites that are drier, more exposed and often more saline. *P. vaginata* was relatively more drought resistant, in that its biomass was only reduced by drought stress at the seedling stage and mortality was low, while the

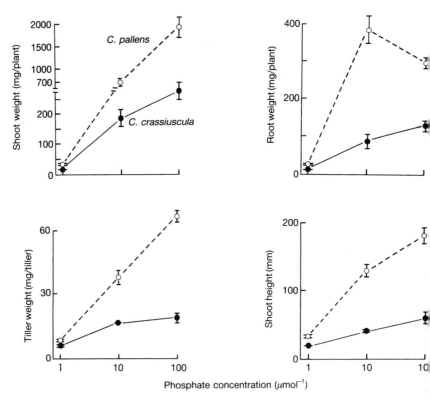

Fig. 6.4. Size characteristics of the seedlings of two alpine grasses, *Chionochloa pallens* and *C. crassiuscula*, grown under different phosphate concentrations. (Simplified from Chapin *et al.* 1982.)

biomass of *P. algida* was reduced at all ages, and mortality was high under stressful conditions. *P. vaginata* had a slower growth rate and showed a lower carbon allocation to reproduction than *P. algida*, but tended to live longer. Many other comparative studies covering a range of stressful environments have illustrated the association between stress resistance and slow growth rate.

Growth rate depends on many variables, including respiration and photosynthetic rates, and the allocation of energy to growth versus reproduction. Characteristics such as storage tissue, extensive root systems, salt glands, and thick cuticles, which enable plants to survive in extreme environments, cost energy to produce and maintain, and it is these characters that may be of primary importance for survival in a stressful environment. Diversion of energy from growth to the production of such systems may decrease growth rate, making slow growth a consequence of stress resistance. If the resistance mechanism is expressed constitutively, such energetic costs will persist when plants are grown under optimal conditions. This raises the question of whether slow growth rate is an adaptation enabling the conservation of resources or a consequence of stress resistance responses that utilize metabolic energy.

Grime (1979) and Chapin (1980) suggest several possible advantages of slow growth that are related to resource conservation, particularly with respect to nutritional stress. One possibility is that slow-growing species are less likely to exhaust resources, while plants with high growth rates have high intrinsic respiration rates that cannot be supported in habitats where resources are scarce. This argument requires that adjacent individuals do not compete for nutrients or other resources, because slow growers would be outcompeted for the few resources that are available. The persistence of slow growth as a resistance mechanism would otherwise require group selection to prevent invasion of a population by a genotype that promotes rapid growth.

A second possibility is that slow growers may function close to their optimum photosynthetic and metabolic rates under conditions of limited resources, giving them an advantage over fast-growing individuals, which function at sub-optimal levels under the same conditions. This argument assumes a trade-off between optimal performance under stressful conditions and optimal performance under benign conditions, discussed further in Chapter 7. Low light habitats are an example of a stressful situation where competitors or ruderals may not function optimally, and where low respiration rates may be an advantage. Mahmoud and Grime (1974) examined growth of shade-resistant *Deschampsia flexuosa*, and shade-sensitive *Agrostis tenuis* and *Festuca ovina*. The three species had the same light compensation points, where the energy produced by photosynthesis was balanced by the energy lost via respiration. However, the shade-resistant species was able to maintain a higher relative growth rate in poor light because of a lower respiration rate, which conserved the energy obtained from photosynthesis. Associations between low respiration rate and shade resistance have also been found in tree seedlings (Grime 1965; Loach 1967).

A third advantage of slow growth is that plants may obtain resources in excess of their immediate requirements, which could enable them to continue growth after exhaustion of resources in the environment. Slow growers, including many Australian plants adapted to low-nutrient habitats, can continue growth with little absorption of nutrients from the soil. Barrow (1977) showed that *Banksia grandis* used a phosphorus reserve in seeds to enable it to continue to grow at its maximum rate without phosphate for 22 weeks. It thus attained a higher biomass than other species that normally grew faster in the presence of phosphate.

Finally, slow growth may enable plants to take advantage of an intermittent resource supply. Crick and Grime (1987) showed that a fast-growing *Agrostis stolonifera* had relatively higher rates of nitrogen capture and dry matter production than slow-growing *Scirpus sylvaticus*. *Agrostis stolonifera* has a small root system which could respond to areas of high nutrient concentration (that is, high phenotypic plasticity), but which was inefficient in the utilization of short pulses of nutrients. In contrast, *S. sylvaticus* maintained a large and unresponsive root mass that did not respond to localized areas of nutrient enrichment. However this root system appeared to be more successful than that of *A. stolonifera* in capturing nutrients from short pulses.

There is some evidence for an association between slow growth and stress resistance at the intraspecific level, as illustrated by the following examples, most of which were mentioned in Chapter 4. A *Spartina patens* population originating from a sand dune where they were subjected to drought, salt, and nutrient stress showed slower growth rates under protected, high nutrient conditions than populations from a swale or marsh (Silander and Antonovics 1979). In the Park Grass experiment where *Anthoxanthum odoratum* was maintained in a patchy environment for 60 years, plants originating from low yield areas had slower growth rates under favourable conditions than plants originating from high yield areas (Davies and Snaydon 1976). Differences in growth rate evolved despite considerable gene flow between plots. Survival of varieties of the tall fescue (*Festuca arundinacea*) from different geographic locations at low temperatures was negatively correlated with growth rate (Robson and Jewiss 1968). Respiration rates of *Oxyria digyna* populations from northern latitudes were higher than those of southern alpine populations over a range of temperatures, and the northern populations were less resistant to high temperature stress than southern populations (Mooney and Billings 1961).

One implication of the association between slow growth and stress resistance is that plants exhibiting slow growth in an environment may not necessarily be poorly adapted to that environment. Genecological studies which involve reciprocal transplants between locations provide examples where plants appear to perform more poorly in their site of origin. For example, Antonovics and Primack (1982) described several cases where *Plantago lanceolata* originating from one site showed a slower growth rate at its site of origin than plants originating from other locations. Slowly growing plants that have a long term advantage in a stressed environment may appear to be poorly adapted to that environment in short term experiments. Assessing the relative fitness of populations requires studies that span several generations, because slow growth rate may be an advantage under intermittent stress conditions. For example, Menges and Waller (1983) found that slow-growing perennial sedges and grasses predominated in areas exposed to intermittent stress of flooding, while fast-growing herbaceous species predominated in flood-free areas.

In summary, there is evidence at the interspecific level that slow growth is a feature of plants from stressful habitats and an association between slow growth and stress resistance may also hold at the intraspecific level. Slow growth may be a general adaptation to unfavourable conditions which enables plants to conserve limited resources. However, it is also possible that some cases of slow growth are a consequence of resistance mechanisms which are specific to some environmental stresses. Most of the emphasis on the potential advantages of slow growth has been on soil nutrients, although slow growth may also help in coping with other limiting resources (Grime 1979; Mahmoud and Grime 1974), particularly as it is a feature of plant species from a diversity of stressful habitats. Additional experiments are required to determine whether slow-growing plants that are selected under one environmental stress have increased resistance to other environmental stresses. Such experiments may involve taking lines that have been selected in a low-nutrient environment and growing them under low-moisture conditions to indicate whether slow growth is a mechanism leading to increased general stress resistance.

6.4 General stress resistance: other mechanisms

Most research on resistance mechanisms has focused on the elaborate morphological and physiological adaptations of species to habitats with continuously extreme environmental conditions. This has emphasized the specific nature of many resistance mechanisms. However resource conservation is only one of the possible mechanisms that could provide resistance to multiple environmental stresses. Other general mechanisms are suggested by similarities in the way in which different stresses cause damage and similarities in the physiological reaction of organisms to different stresses.

6.4.1 Interactions between environmental stresses in plants

Resistance to different stresses may be correlated if the exposure of an organism to one environmental stress induces several other types of secondary stresses. In plants, Levitt (1980) has considered such interactions between stresses by distinguishing between stress effects at the primary, secondary, and tertiary levels. The stress itself is responsible for injuries at the primary level; injuries may also be caused by a secondary stress that is a consequence of the primary stress, such as high temperatures and freezing causing a water stress. Different secondary stresses may in turn trigger a similar tertiary stress. Stress responses can therefore involve the same mechanism since different primary stresses cause similar secondary or tertiary stresses. The relationships between the three levels of stress effects are given in Table 6.4, and illustrate, for example, that water stress can arise from seven primary stresses and nutrition stress can arise from four primary stresses. There is obviously considerable potential in plants for countering different environmental stresses by mechanisms providing resistance to one stress.

As in the case of resource conservation arising through a reduction in metabolic rate, other mechanisms providing resistance to different stresses may share similarities (Levitt 1980; Fig. 6.5). Levitt separated common stress responses based on interactions between the primary, secondary and tertiary levels, as discussed above. The next division separates out resistance associations from parallel selection, where plants have been selected for increased resistance to stresses that tend to occur together, such as heat and dryness in arid environments. Plants may show high or low resistance to both heat and desiccation, but the association between these stresses may not have a genetic basis. Common resistance mechanisms can be split into whether or not the strain resulting from a stressor is encountered, and, if encountered, whether it is tolerated or avoided. In general, tolerance mechanisms are expected to be more closely correlated in plants than avoidance mechanisms. Heat, freezing, and drought tolerance may be correlated because they result in dehydration. Many stresses lead to membrane damage (Levitt 1980), giving the possibility of a common resistance mechanism, although different types of membrane damage can arise from different stresses. Resistance mechanisms may also be correlated if

Table 6.4. Interactions between stresses in plants. Injury may be caused by an environmental stress (primary stress), which in turn can lead to other types of stresses (secondary and tertiary) being exerted on a plant. The secondary stresses may be more or less injurious than the primary stresses. (After Levitt 1980.)

Primary stress	Secondary stress induced by primary stress	Tertiary stress induced by secondary stress
Primary stress as major cause of injury		
Chilling low temperature	Water deficit	
Freezing (intracellular)		
High temperature	Water deficit	Mineral deficit
Water deficit	Mineral deficit	
Radiation (UV, ionizing)	High temperature	Water deficit
Ultrasound	High temperature	
Secondary stress as major cause of injury		
Freezing (extracellular)	Water deficit	
Radiation (visible, infrared)	High temperature	
Flooding	Oxygen deficit	Mineral deficit
Pressure	Oxygen excess	
Wind	Water deficit	
Primary and secondary stresses equally damaging		
Salt ion	Water deficit	Mineral deficit

they result in similar metabolic disturbances, or share common repair mechanisms.

While these considerations indicate the possible existence of generalized stress resistance mechanisms, it should be emphasized that stress resistance is often expected to be associated with specific mechanisms. Responses to the same stress can also involve different mechanisms: resistance to a stress may occur through avoidance or tolerance, and by different avoidance or tolerance mechanisms. For example, salt-resistant glycophytes avoid the stress of a high intracellular NaCl concentration by actively excluding salt. On the other hand, halophytes show resistance to environmental salt in the presence of increased intracellular NaCl levels (Greenway and Munns 1980). These high intracellular salt concentrations overcome the problem of a water stress and are compartmentalized in vacuoles away from the cytoplasm, allowing tolerance. Salt resistance may generally come about by a range of mechanisms, involving different genes (Cheeseman 1988). There are also several ways in which plants can respond to a drought stress. Drought stress may

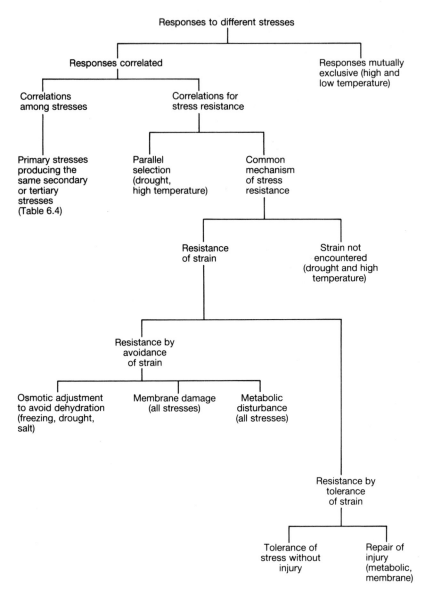

Fig. 6.5. Possible associations between responses to different environmental stresses in plants. For further explanation, see text. (Modified from Levitt 1980.)

be avoided by a high water-absorbing ability which allows for a high transpiration rate; water conservation, another drought avoidance mechanism, is incompatible with a high transpiration rate. These diverse mechanisms can be found in different species growing in the same region (e.g. Lo Gullo and Salleo 1988). Drought resistance in crop plants can also occur via increases of turgor pressure because of changes in the root system or changes in stomatal function, altered levels of epicuticular lipids on the leaf surface, and increased proline accumulation (Blum 1988).

Resistance mechanisms, for example, resistance to high and low temperature, may be mutually exclusive rather than correlated. Plants that are native to cool environments often attain maximum photosynthesis at low temperatures, while plants from hot environments often attain maximum photosynthesis at high temperatures and perform poorly at low temperatures (Berry 1975). This suggests a trade-off such that a plant cannot perform well under both temperature extremes, which is consistent with differences in mechanisms of heat and cold temperature resistance. As discussed in Levitt (1980), proteins most readily denatured at high temperatures are those with a low proportion of hydrophobic bonds, and these are likely to have greatest stability at low temperatures. Chill-resistant plants have higher levels of lipid unsaturation, while a lower level of unsaturation is found in heat-resistant plants. In general, increased resistance to a stress is expected to be associated with decreased resistance to another stress when the stresses represent opposing extremes of the same environmental factor.

Nevertheless, there is much non-genetic evidence to suggest that resistance to different stresses often has a common physiological basis in plants (Levitt 1980). In particular, exposure to a sub-lethal dose of a stress can lead to increased resistance to other stresses (cross-resistance). For example, exposure to chilling temperatures can confer resistance to chilling, freezing, drought, salt, and heat.

Although the above discussion has focused on plants, cross-resistance in animals also indicates common resistance mechanisms. In mammalian cells, heat exposure and ethanol induce thermal resistance as well as resistance to the drug adriamycin, and this is probably related to common changes in membrane fluidity (Li and Hahn 1978). At the whole organism level there are several examples in mammals where exposure to one environmental stress has been shown to increase resistance to another environmental stress (Hale 1970). In fish, the use of reference toxins has been advocated for measuring the impact of prior stressors (Wedemeyer et al. 1984). Reduced resistance to a reference toxin often indicates that individuals have been exposed to a prior stressor, suggesting that the effects of different stresses are often cumulative. This implies that increased resistance to one stress will increase resistance to another stress when different stresses are encountered by the same organism.

6.4.2 Stress proteins, stress metabolites, and other inducible responses

Heat-shock proteins could form the basis of a general resistance mechanism in plants and animals. Although these proteins are triggered by a large number of environmental and artificial stresses (Section 3.2), it is not known if they increase resistance to all of the stresses that induce them. The evidence for increased stress resistance is often only indirect and has not extended much beyond a consideration of thermal resistance. Experiments where oocytes are injected with denatured protein suggest that heat-shock proteins may be produced in response to protein denaturation (Ananthan *et al.* 1986); they should, therefore, have the potential for providing a broad resistance to any stress that results in protein denaturation. In *Drosophila melanogaster* heat-shock proteins are triggered by cold treatment, and increase resistance to cold stress (Burton *et al.* 1988).

Heat-shock proteins are not the only molecules that are influenced by a range of environmental stresses. The concentration of molecules that regulate a cell's water potential may change in response to diverse stresses. For example, proline accumulates in response to salinity, drought, low temperature, ozone, and SO_2 (Le Redulier *et al.* 1984). Such molecules are important in maintaining a cell's osmotic balance which can be upset by salinity and drought stresses, but their role in resistance to other stresses is not known. Genetic variation in proline accumulation has been described in crop plants and has been associated with drought resistance (Blum 1988). Similarly, the disaccharide trehalose accumulates in yeast in response to several stressors such as ethanol, copper sulphate, starvation, and hydrogen peroxide (Attfield 1987). This compound may function as a protectant during freezing or as a membrane protectant during desiccation.

Finally, chemicals known as 'stress metabolites', including growth hormones such as abscissic acid and ethylene, are commonly associated with diverse stresses in plants (reviewed in Ayres 1984). For example, ethylene is produced by plants in response to mechanical wounding, temperature extremes, radiation, chemicals, drought and infection. The function of this increase in hormone production is not known, although ethylene may regulate the synthesis of phytoalexins which can be toxic to infectious agents and may provide resistance to biotic factors. The production of ethylene may form the basis for a mechanism of stress damage that is common to various atmospheric pollutants. Mehlhorn and Wellburn (1987) showed that ethylene production by peas makes the plants susceptible to ozone, and the increased production of ethylene in response to other pollutants enhances ozone-mediated leaf injury. Stress metabolites may therefore represent a common form of damage from diverse stresses as well as reflecting resistance mechanisms. Genetic variation in levels of these compounds is likely to influence stress resistance, as suggested by the association between abscissic acid and drought resistance in crops (Blum 1988).

6.4.3 General adaptation syndrome

In mammals, the concept of a general resistance mechanism has been developed by Selye (see Section 1.1), based on his initial observations of nonspecific stress effects on the mammalian endocrine system (Selye 1950, 1952, 1956). Concentrations of hormones alter in a specific manner in response to a range of environmental stresses and other inputs; these form a component of Selye's 'general adaptation syndrome' which comprises three stages. In the first (stage of alarm reaction), the stress factor causes a release of adrenocorticotrophic hormone from the pituitary gland, which incites cells of the adrenal cortex to discharge corticoid hormones. In the second stage, the organism acquires resistance to a stress (stage of resistance) and the adrenal cortex becomes laden with fat droplets. These droplets are lost in the third stage (stage of exhaustion), when stress resistance is lost. Organisms may reach the last stage only when a stress persists. In rats these stages have been found to be initiated by a wide range of environmental and biotic changes, including temperature extremes, drugs, infection and forced muscular work. Selye (1956) argued that the physiological and morphological changes reflect a mechanism of general stress resistance rather than just a general reaction to stress, although there is not much empirical evidence to distinguish between these alternatives (Munday 1961).

A similar general adaptation syndrome may occur in marine animals (Wedemeyer et al. 1984; Koehn and Bayne 1989) where primary hormone effects, secondary hormone-induced changes, and tertiary whole-animal effects can be recognized. In fish, similar physiological changes may occur in response to handling, disease, fright, forced swimming, anaesthesia, scale loss, or rapid temperature changes (Wedemeyer et al. 1984).

The common hormonal changes associated with these generalized stress responses suggest that variation related to the endocrine system may form the basis for variation in general stress resistance, but there is little information available. In one of the few genetic studies of hormones and environmental stress, Jonsson et al. (1988) examined levels of hydrocorticosterone in mice selected for increased and decreased resistance to heat. Animals from the stress resistant line had a higher concentration of this hormone in their blood plasma than animals from the susceptible line. More experiments of this type relating hormone levels to general stress resistance are required, in both vertebrates and invertebrates.

6.4.4 *Concluding remarks*

Several mechanisms may influence resistance to more than one type of environmental stress. It is expected, therefore, that increased resistance to one stress may often be accompanied by increased resistance to another stress, or by a decrease in some cases. For example, a consideration of resistance mechanisms suggest that heat and desiccation resistance might often be correlated, while heat and cold resistance or light and shade resistance in plants might often be independent or negatively correlated. These considerations can be used to predict the nature of associations between resistance traits at the genetic level.

6.5 Genetic correlations: origin and measurement

Before discussing associations between stress resistance traits at the genetic level, we will briefly comment on genetic correlations that are used to quantify such associations. Genetic correlations are also central to much of the discussion in Chapter 7.

Correlations between traits can arise from the pleiotropic nature of mutations. For example, the above discussion suggests that genes decreasing metabolic rate may increase resistance to a number of stresses, and may therefore have pleiotropic effects on different resistance traits which could result in positive correlations. Genetic correlations may also arise because of linkage: a positive genetic correlation between two traits may occur when an allele increasing the expression of one trait is non-randomly associated in gametes with an allele at a different locus which increases expression of another trait. This means that non-random associations can persist because of linkage between loci, especially when linkage is tight.

Genetic correlations may also depend on past selection for traits. Falconer (1981) argued that joint selection for the increased expression of two traits can generate a negative genetic correlation between them. Genes with pleiotropic effects on both traits in the direction of selection will be rapidly fixed and those with undesirable effects on both will be rapidly lost. However, genes with desirable effects on one trait and undesirable effects on the other trait (i.e. genes showing antagonistic pleiotropy) will be left segregating in the population. This residual genetic variation will generate a negative genetic correlation between the traits which persists within a population. Continuous selection for increased resistance to two stresses could therefore lead to the fixation of genes that increase resistance to both stresses. The remaining genetic variation results in a negative correlation between the stresses, even when much of the resistance to these stresses is associated with a common mechanism.

Genetic correlations may also be influenced by the intensity of selection. As discussed in Section 5.6, major mutations will tend to be favoured when the selection intensity is high. Major mutations are more likely to be associated with deleterious fitness effects than are minor mutations (Section 1.2). A negative genetic correlation between stress resistance and fitness-related traits may therefore arise when the selection response to an environmental stress is based on major mutations, whereas negative correlations may be weaker or absent when minor mutations form the basis of a selection response.

Finally, genetic correlations depend on the environment in which they are measured. If different genes affect a character in two environments, then the genetic correlation between this character and other traits can change as the environment changes. It is possible for one set of genes to influence resistance at low stress levels, while another set of genes influences resistance at high stress levels. For example, a non-specific mechanism of heavy metal resistance in plants can provide generalized resistance at low stress levels, while metal-specific mechanisms are important at high stress levels (Symeonidis *et al.* 1985). Genetic correlations between resistance to different heavy metals will therefore become less positive as the stress level is increased. Correlations between traits can also change across environments, when the same genes influence the traits but their effects depend on the environment. Plant productivity in a stressful environment may be improved by genes increasing stress resistance, but the same genes could decrease productivity in a benign environment because of energetic costs and other costs associated with producing a resistance mechanism. The genetic correlation between productivity and stress resistance could therefore switch across environments, as will be discussed further in Chapter 7.

Several standard designs described in texts on quantitative genetics (e.g. Falconer 1981) can be used to measure genetic correlations. A common approach is to artificially select one trait and look for changes in other traits after selection. Clark (1987) has pointed out two problems with this approach: the first is inadvertent direct selection on traits scored as correlated characters. For example, lines of insects selected for desiccation resistance could inadvertently be selected for starvation resistance because of the length of time they are left without food during the desiccation process, leading to apparent positive or negative genetic correlations when none exist. This can be overcome by carefully maintaining control lines that are treated like the selected lines except that they are not exposed to the selection pressure. The second problem is that the environment where the lines are maintained during selection may differ from the environment to which the base population is adapted. Genetic changes associated with adaptation to the new environment may occur while lines are being selected and these changes are not related to the selected trait. This problem can be important when individuals adapted to field conditions are brought into a laboratory and selection is started before the population has adapted to the new environment. Another limitation of many selection experiments is that selection and control lines are not independently replicated. This makes it difficult to separate correlated responses from random genetic changes due to genetic drift or mutation.

Another technique for characterizing genetic correlations is to compare parents and their offspring or to compare siblings from defined crosses (usually a half-sib design). In a parent–offspring comparison, traits are scored on parents as well as their offspring, and genetic correlations are obtained from variance and covariance estimates of the traits. This design is not particularly useful for stress-related traits because it is difficult to score the resistance of the same parental individual to different environmental stresses, particularly when stress measures involve mortality. In a half-sib design, several males are each mated to several females and progeny from each family are scored for the traits being correlated. In measures of resistance involving mortality, some family members can be exposed to one stress, and other sibs can be exposed to a different stress. The male and female contributions to the variance and covariance between traits provide an estimate of the additive genetic correlation. These estimates tend to have large standard errors because the genetic correlations are functions of covariance components which are themselves estimated with limited precision. A large number of families therefore have to be tested to obtain meaningful correlation estimates.

Finally, genetic correlations may be obtained by characterizing traits in a series of inbred or partially inbred lines and examining correlations between line means. Rose (1984) pointed out that inbreeding is a problem in examining genetic correlations among traits related to fitness because it can lead to the fixation of deleterious recessive alleles in some lines. These alleles are often associated with poor performance for a range of fitness-related traits and will therefore tend to generate positive correlations between traits even though they will not contribute much to the genetic correlation in outbred populations. An additional problem with using partially inbred lines, such as isofemale strains, is that correlations among line means can include dominance effects. These do not contribute to additive genetic correlations which determine the extent to which genetic changes in one trait will be associated with correlated changes in another trait. Genetic correlations have also been obtained by extracting intact chromosomes carrying inversions that suppress crossing over, and characterizing correlations between lines that are isogenic for different chromosomes. Unfortunately, isogenic lines may represent an artificial genetic situation because they do not occur in nature. This technique is also restricted to the few species where inversions that supress crossing over have been characterized.

This brief discussion emphasizes the fact that genetic correlations have to be interpreted cautiously because of the number of factors that can influence them, and because of the difficulties in their measurement. Genetic correlations should be considered in conjunction with estimates of genetic variation for the traits being correlated. A large positive correlation between two traits may be meaningless if heritabilities are low, because a highly correlated trait with a low heritability may change only slightly during selection. The size of the genetic correlation should also be emphasized rather than just its sign. A large positive or negative correlation may indicate a common physiological basis for two traits, and suggests that the traits will usually tend to vary together. However a correlation that is significantly less than 1 or greater than −1 indicates that traits can be altered independently to some extent.

6.6 Genetic correlations among stress resistance traits

Few of the early *Drosophila* experiments on environmental stress resistance considered correlations between resistance traits. An exception is the work by Matheson and Parsons (1973) discussed in Section 6.2. Resistance to CO_2 and N_2 in 18 *D. melanogaster* isofemale strains was found to be highly correlated, with correlation coefficients of 0.62 for females and 0.74 for males. Since both stresses lead to anoxia, resistance via a reduction in oxygen consumption (metabolic rate) probably provides a physiological basis for this correlation. In a recent study, Sierra *et al.* (1989) selected for increased resistance to acrolein in *D. melanogaster* and found a correlation with CO_2 resistance, which is probably also associated with underlying variation in metabolic rate.

Turning to other *Drosophila* studies, Cohan and Hoffmann (1986, 1989) selected lines from four *Drosophila* species for resistance to knockdown by ethanol fumes and tested lines for other traits related to stress resistance. Increased knockdown resistance was generally related to other measures of ethanol resistance, including larval viability in medium with ethanol, and adult longevity in the presence of ethanol fumes. In addition, the selected *D. melanogaster* lines were more resistant than controls to toxic concentrations of acetaldehyde and acetone, although they did not differ for starvation resistance (Cohan and Hoffmann 1986). The same correlated responses were observed when knockdown resistance was selected in *D. simulans* (a sibling species of *D. melanogaster*). In *D. pseudoobscura*, other measures of ethanol resistance and acetaldehyde resistance did not increase in selected lines, although knockdown resistance to ethanol increased less in this species than in the other species (Hoffmann and Cohan 1987). In contrast, increased knockdown resistance in *D. persimilis* (a sibling species of *D. pseudoobscura*) was associated with increased resistance to starvation and acetaldehyde, but there was no correlated response for resistance to acetone (Cohan and Hoffmann 1989). Selection in these experiments was carried out on replicated lines which showed consistent correlated responses. These results provide some evidence for non-specific increases in stress resistance when selection is carried out for a chemical stress, although the correlated responses differed between species. There was also evidence that conspecific populations of *D. melanogaster* and *D. pseudoobscura* showed different correlated responses. For example, acetone resistance showed a correlated increase in both replicate lines set up from some *D. pseudoobscura* populations and decreased or did not change in lines from other populations (Table 6.5). These results indicate that genetic correlations among stress resistance traits can vary between conspecific populations and closely related species.

We have examined correlated responses for stress resistance traits in *D. melanogaster* lines selected for increased resistance to desiccation

Table 6.5. Mean acetone resistance (% survival) of *Drosophila pseudoobscura* lines from conspecific populations selected for increased knockdown resistance to ethanol. Standard deviations are given in brackets. (From Hoffmann and Cohan 1987.)

Population	Control lines		Selected lines	
	1	2	1	2
OX	76 (13)	58 (3)	14 (4)	29 (17)
WA	49 (18)	32 (17)	52 (20)	63 (12)
CA	14 (7)	10 (14)	19 (14)	5 (6)
KE	28 (17)	45 (13)	57 (11)	29 (14)
PC	44 (17)	39 (20)	78 (26)	71 (23)

(Hoffmann and Parsons 1989a,b). The three selected lines showed increased resistance to starvation, toxic concentrations of ethanol and acetic acid, heat shock, and gamma radiation compared to three control lines; most of the differences between the selected and control lines were large (Table 6.6). There was no correlated response for resistance to acetone, ether or cold shock, but resistance to desiccation, ethanol and starvation were genetically correlated. This was confirmed by tests performed on isofemale strains from the F_2s of a cross between one of the selected lines and one of the control lines. Many of these correlated responses are consistent with the hypothesis that increased stress resistance involves a decrease in metabolic rate (Section 6.2). A decreased metabolic rate is expected to increase resistance to heat stress but not cold stress, and will also increase resistance to starvation. The increased radiation resistance can be reconciled with metabolic rate changes because systems with low metabolic rates (inactive tissue, quiescent stages, mature adults) are more resistant to radiation damage than systems with high metabolic rates (active tissue, developing individuals) (Grosch and Hopwood 1979). However the possibility remains that other mechanisms, such as changes in the lipid components of membranes, provided general stress resistance and influenced metabolic rate.

Turning to other animals, Bradley (1981) found positive correlations between heat and cold resistance in the copepod *Eurytemora affinis*. The correlations were based on family means and were weak, the coefficient for males being 0.13 and for females 0.11. White *et al.* (1970) carried out selection experiments for genetic adaptation to temperature extremes in a hymenopterous parasite (*Aphytis lingnanensis*). There was a small correlated increase in resistance to high temperature in the line selected for low temperature resistance, but independent selection lines were not

Table 6.6. Correlated responses to selection in *Drosophila melanogaster* lines selected for increased resistance to desiccation. (Data from Hoffmann and Parsons 1989a,b.)

Stress	Mean (selected)	Mean (control)	Selected/control
15% ethanol*	62.9	27.5	2.29
Starvation*	98.2	78.6	1.25
15% acetic acid*	22.9	12.5	1.83
Heat†	70	45	1.55
Radiation‡	64	25	2.56

*Time (hours) taken for 50% of flies to die.
†% flies upright after 1.5 h at 37°C.
‡% flies alive 22 h after a dose of 1.2 kGy.

maintained. In contrast, Jonsson *et al.* (1988) found that mice selected for increased heat resistance were less resistant to a cold period than mice selected for decreased heat resistance. This correlated response may have been mediated by body size because mice from the heat-resistant line were relatively smaller.

Comparisons of varieties of plants provide several examples where resistance to different stresses is correlated (Levitt 1980). In an early study, Coffmann (1957) found that cold resistance, measured by winter survival, and heat resistance, measured by a brief exposure to a high temperature, were associated in oat varieties. Of the 12 varieties tested, two that showed extremely low winter survival were also the most susceptible to a heat stress. The remaining varieties tended to show high or intermediate levels of resistance to both stresses. McKersie and Hunt (1987) measured resistance to ice encasement, low temperature flooding, and freezing in 34 strains of winter wheat by scoring survival after stress exposure in plants from four harvest times, representing two different times of seeding and two sampling dates (spring, autumn). Overall differences between strains were found for all traits, although there was also a significant strain by environment (harvest time) interaction for icing and freezing, indicating that resistance differences between strains varied somewhat with environmental conditions. Scores for each strain were averaged over the environments and correlations between resistance traits were determined from the strain means. All correlation coefficients were positive and large (0.72 to 0.75), suggesting that similar mechanisms controlled resistance to ice encasement, flooding and freezing.

Genes with pleiotropic effects on different stress resistance traits can indicate common resistance mechanisms that may contribute to genetic correlations between traits. A mutant of the green alga, *Chlorella sorokiniana*, which was originally selected for its ability to grow autotrophically under high O_2 tension was also more resistant to UV irradiation and to an O_2 generating antibiotic, streptonigrin (Pulich 1974). These pleiotropic effects may be due to an altered O_2 detoxification system involving elevated levels of superoxide dismutase. This polymorphic enzyme has been shown to be important in irradiation resistance in *D. melanogaster*: the S form of the enzyme (which exhibited highest *in vitro* activity) gave the greatest protection, and furthermore, the fitness of the high activity S allele was increased in an irradiated population relative to a non-irradiated control (Peng *et al.* 1986). A yeast strain that carried a deletion for the structural gene coding for the precursor of the protein ubiquitin was sensitive to stresses caused by high temperature, starvation and amino acid analogs, suggesting a role for ubiquitin in the heat shock response and general stress resistance

(Finley *et al.* 1987). Mutants for heat-shock proteins may be particularly useful for investigating the role of these proteins in general stress resistance. Mutant strains with altered patterns of heat-shock protein synthesis can be isolated by their susceptibility to thermal shock (e.g. Loomis and Wheeler 1982), and it would be interesting to characterize these strains for responses to other stresses.

Genotypes of polymorphic enzyme loci may also be used in the study of pleiotropic effects. Variation at the structural locus for alcohol dehydrogenase (Adh) in adults and larvae of *D. melanogaster* has been related to temperature and ethanol resistance (Section 3.2; see also Chambers 1988). Genotypic differences in the recovery of adults after a heat shock were only found when they were transferred to medium with ethanol after the shock (Sampsell and Sims 1982). Adh genotypes may vary in their production of stress proteins, either through a change in the concentration of intracellular ethanol required to trigger heat-shock proteins or through an effect on other metabolites (Stephanou and Alahiotis 1986).

The above discussion provides some evidence that resistance to different environmental stresses can be positively correlated at the genetic level. Insufficient research has been carried out to make generalizations, although variations in metabolic rate may underly some of the positive correlations. There is a need for experiments that simultaneously consider a range of stresses, including those related to low metabolic rate and slow growth rate. Resistance mechanisms associated with changes in centralized metabolic processes are more likely to provide general resistance than those associated with environmental filtering and peripheral metabolic pathways. Many interesting research questions have not yet been considered empirically. None of the studies have addressed the relative importance of specific versus general stress resistance by partitioning the genetic variance into these components. Genetic correlations in populations exposed to different levels of stress selection have not been examined to see if a history of directional selection influences correlation among resistance traits.

It should be emphasized that genetic variation for resistance to some stresses, particularly chemical stresses, will probably be specific rather than general. Genes providing resistance to an insecticide may provide cross-resistance to other insecticides, but it is unlikely that these genes will influence resistance to other environmental stresses. The extent of cross-resistance among insecticides will depend on the type of mechanism, because genes affecting the rate of entry of an insecticide may tend to be associated with wider cross-resistance than genes affecting detoxification pathways (Plapp 1976). Turning to other chemicals, Deery and Parsons (1972b) found that variation in ether resistance among *D. melanogaster* isofemale strains was uncorrelated with variation in chloroform resistance. These chemicals did not show a correlated response in lines selected for increased desiccation resistance (Hoffmann and Parsons 1989*b*). Heavy metal resistance in plants tends to be metal-specific with little cross-resistance (Turner 1969). However low levels of heavy metal resistance may be general rather than specific because Symeonidis *et al.* (1985) found that cultivars of *Agrostis capillaris* had low levels of resistance to metals other than those present at toxic levels in soils from which they were derived. In addition, populations of tufted hair grass (*Deschampia cespitosa*) that have invaded polluted areas were resistant to heavy metals not found in the polluted soils, suggesting genetic correlations for resistance to different heavy metals (Hutchinson 1984).

The study of genetic correlations among traits related to stress resistance needs to be extended to the population and species levels. If genetic correlations act as evolutionary constraints as envisaged by Lande (1976), Cheverud (1984) and others, correlations between populations may extend to higher levels and help to explain patterns of interpopulation and interspecific variation. For example, *D. melanogaster* populations on the east coast of Australia show genetically based differences in ethanol resistance (Section 4.2) which decreases towards the tropics (Parsons 1980) and there are parallel clines in ethanol resistance in *D. immigrans* and *D. simulans* (Stanley and Parsons 1981). Genetic variation in the ethanol resistance of *D. melanogaster* shows a similar latitudinal association on the west coast of North America (Cohan and Hoffmann 1986). These patterns are difficult to explain in terms of natural selection because the ethanol levels of tropical and temperate *Drosophila* resources appear to be similar (Gibson *et al.* 1981). However one possible explanation for this geographic pattern is that ethanol resistance was not selected directly but changed indirectly because it was correlated with other traits. For example, desiccation resistance is also lower in the northerly populations of the humid tropics and subtropics of Australia which is predicted by the reduced desiccation stress at these locations (Parsons 1980). Desiccation resistance is correlated with ethanol resistance (Hoffmann and Parsons 1989*a*) and this correlation may account for the clinal variation in ethanol resistance. The validity of this hypothesis will depend on whether the population differences in ethanol resistance are attributable to a general stress resistance mechanism or a mechanism that is specific to ethanol.

6.7 Summary

In animals, low resting metabolic rates may be generally associated with increased resistance to desiccation, heat, starvation, and other stresses. Low metabolic rate conserves energy which may extend longevity under stressful conditions when the potential for energy acquisition is reduced. Most supporting evidence for low metabolic rate as a general resistance mechanism is from interspecific comparisons and plastic responses to stress, but a few genetic studies are consistent with this hypothesis.

In plants, slow inherent growth rate is a common feature of species from stressed environments. There is a link between slow growth rate and increased resistance to low nutrient, and low light conditions, but it is not clear whether slow growth is a cause or an effect of resistance to other stresses. A limited number of intraspecific comparisons support an association between stress resistance and slow growth rate.

Correlations between resistance to different stresses may arise because the stresses cause the same physiological or intracellular stress or similar types of stress damage. Phenotypic changes in heat shock proteins, hormones, and the composition of membrane lipids may alter general stress resistance.

Few genetic studies have examined correlations between resistance to different stresses, although there is evidence in *Drosophila* that selection for increased resistance to one stress reduces metabolic rate and increases resistance to a range of other stresses. Genetic correlations among stress resistance traits can vary in different populations and closely related species, but may be predictable to some extent in situations where the underlying resistance mechanisms are understood.

7. Stress response, costs, and trade-offs

From the discussion already presented, it is clear that individual differences in performance under stressful conditions will normally have a genetic basis. We now consider costs associated with increased stress resistance or evasion as expressed in a non-stressful environment. In other words, is there a trade-off between performance under stressful conditions and performance under other environmental conditions such that individuals that perform well under stress perform poorly in optimal environments? We start by discussing the physiological costs of stress resistance, then examine phenotypic variation in resistance and stress evasion. The evidence for trade-offs at the genetic level will be examined by considering specific genotypes and quantitative genetic variation. This includes a discussion of the association between genetic variation in stress resistance and life-history traits, and a consideration of the effects of stress on correlations among life history traits.

7.1 Costs of stress resistance

Costs associated with stress resistance can be considered in terms of energetics and resource allocation. For example, Odum (1985) and Sibly and Calow (1986) treat stress resistance as a component of the energy allocated for maintenance functions. The presence of an environmental stress increases maintenance costs because an organism needs to expend energy to counter its effect. Many cases have been discussed where stressed organisms expend energy to resist damage and maintain a constant internal environment. Active physiological processes requiring ATP or other high energy compounds are involved in processes such as transport against concentration gradients, cell division, and protein synthesis for repair. The increased energy demand may be associated with an increased rate of metabolism, detectable in animals as an increase in oxygen uptake. This metabolic cost can be substantial: Davis *et al.* (1988) found a 1.9-fold increase in metabolic rate in sea otters exposed to crude oil, while Johns and Miller (1982) calculated that maintenance costs in the crustacean, *Cancer irroratus*, were doubled when larvae were exposed to copper and cadmium.

The immediate increase in energy expenditure required for maintenance functions during stress is a plastic response. However, trade-offs may arise when some individuals expend more energy than others on maintenance functions because they possess resistance mechanisms that are not plastic and have to be produced and maintained. The allocation of energy to stress resistance increases survival, but necessarily leaves less energy for reproduction, growth, and other processes. Differences in allocation patterns between individuals result in a trade-off between maintenance (measured by survival) and production (measured by developmental rates and fecundity), in that individuals expending more energy for maintenance have less for production and vice versa.

Such mechanisms will result in energetic trade-offs between stress resistance and fitness under non-stressful conditions. Trade-offs at the interspecific level involving specific resistance mechanisms can be illustrated by some plant examples. There is a trade-off between drought resistance and photosynthesis that affects leaf morphology (Orians and Solbrig 1977). Xeric plants produce metabolically costly leaves with thick cuticles, resin layers, sunken stomata, and associated stem conducting systems to minimize transpiration rates. In contrast, mesic leaves cost less energy to produce and lead to an enhanced growth rate under high moisture conditions. Another example involves the trade-off between phenotypes favoured under high and low light intensity (Mooney and Gulmon 1979). Shade plants generally have lower photosynthetic rates than sun plants at light saturation because of lower levels of photosynthetic enzymes. Mooney and Gulmon proposed that there is insufficient light energy in shade habitats to utilize higher enzyme levels efficiently, so that investment in additional enzyme does not result in additional carbon gain. Such considerations imply a cost for shade leaves under high light conditions and vice versa for sun plants. As a final example, plants may acquire resistance to salt because of the accumulation of solutes, and this causes a marked decrease in the growth of resistant plants under optimal conditions because of energy expended in the accumulation of solutes (Yeo 1983).

An understanding of trade-offs based on energy allocation does not necessarily require the identification of specific stress response mechanisms. Sibly and Calow (1989) used an optimality model to predict the response of animals to environmental stress. They started with a trade-off between growth and mortality, and considered the effects of two types of stresses on this trade-off, one influencing mortality and the other influencing growth. They showed how these types of stresses determine the patterns of investment of organisms in growth and other processes. For example, organisms that are exposed to a poor environment for growth or a high mortality should invest less in defence and repair, and more in production.

Although energetic considerations provide a general way of looking at trade-offs, they only give a partial picture of costs associated with stress response mechanisms. The presence of a resistance mechanism may entail a cost separate from the energy required to produce it, as illustrated by the following examples. In birds, Lima (1986) showed that the amount of fat accumulated in the autumn is a trade-off between the probability of inclement weather (against which it is a defense) and the loss of agility (manoeuvrability and speed) which make birds less able to escape predator attack. Plants capable of high photosynthetic rates under high soil moisture are prevented from extracting moisture from dry soils and vice versa (Orians and Solbrig 1977): high rates of photosynthesis require rapid rates of carbon dioxide entry through the stomata, but this leads to high rates of water loss. Plants from xeric environments may therefore have lower photosynthetic rates and grow more slowly under optimal moisture conditions than plants from mesic environments, because of structural constraints imposed by the stomata. Plants using crassulacean acid metabolism during photosynthesis reduce water loss by limiting gaseous exchange to humid nocturnal conditions, but this restricts the amount of CO_2 available for photosynthesis during the day (Ting and Duggar 1968). Lastly, strains of *E. coli* with a multicopy plasmid carrying the Tn10 tetracycline resistance determinant show severe inhibition of growth because of a physiological interaction with the resistance protein (Moyed *et al.* 1983). The presence of a resistance mechanism in these examples imposes a constraint on other cellular and physiological processes.

Predictions of trade-offs based solely on resource allocation may also ignore trade-offs based on resource conservation. Resistance mechanisms associated with resource conservation, such as decreased metabolic rate and decreased growth rate, may not increase energy expenditure or impose constraints but may nevertheless be associated with reduced fitness under optimal conditions. Animals with a low active and resting metabolic rate may be at a disadvantage when resources are not limiting because they may have lower rates of foraging, reduced mating success, and slower anti-predator responses and development times than animals with higher metabolic rates. The lower growth rate of stress resistant plants may result in reduced reproduction and competitive ability under conditions when resources are not limiting. The anticipated performance of plants originating from productive, stable habitats (competitive phenotype), disturbed habitats (ruderal phenotype) and continuously unproductive habitats (stress-resistant phenotype) is summarized in Table 7.1 (simplified from Grime *et al.* 1986). Slow growing plants are expected to be outcompeted in productive habitats by fast-growing competitors, while the reverse should occur in unproductive habitats as the fast-growing species exhaust nutrient

Table 7.1. Predicted response to stress of competitive, stress-resistant and ruderal plants and ecological consequences in two types of habitat. (Simplified from Grime *et al.* 1986.)

Phenotype	Response to stress	Ecological consequences	
		Productive habitats	Stressed habitats
Competitive	Large, rapid changes in quantity, distribution and morphology of leaves and roots	High rates of water and nutrient uptake to succeed in competition	Tendency to exhaust water and nutrient reserves and increased susceptibility to fungi
Stress-resistant	Changes in morphology slow and small	Overgrown by competitors	Conservative utilization of water and nutrient reserves permits survival of stress periods
Ruderal	Rapid reduction of vegetative growth and diversion of resources into reproduction	Overgrown by competitors	Low reproduction fails to compensate for high mortality

supplies and other resources. Ruderal species tend to be short lived, but delay reproduction in unproductive conditions and this may lead to a failure to reproduce under continuously stressful conditions. This scheme suggests trade-offs between competitive ability and resistance as well as trade-offs between resistance and stress evasion by rapid reproduction.

Increased stress resistance may not always lead to a trade-off between performance under optimal and stressful conditions. Koehn and Bayne (1989) argued that stress resistance will be associated with the efficient use of metabolic resources. Individuals with a high level of efficiency can maintain growth and reproduction under a wider range of environmental conditions than individuals with a lower level of efficiency because they can continue to grow and reproduce when resources become limiting. Having an efficient metabolism may be less important under optimal conditions, but this does not mean that efficiency will be selected against under optimal conditions. We return to the association between stress resistance and metabolic efficiency below when discussing the association between growth rate and heterozygosity.

In summary, physiological considerations suggest that trade-offs between stress responses and performance under optimal conditions may arise from patterns of resource allocation, constraints imposed by resistance mechanisms and resource conservation. The evidence that these factors lead to trade-offs at the genetic level is discussed below.

7.2 Fitness costs for genotypes resistant to chemical stresses

We start by considering genotypes that are associated with resistance to chemical stresses. Do resistant genotypes show decreased fitness in the absence of a stress? We might expect deleterious fitness effects on the basis of energy expended in the resistance mechanism (e.g. keeping toxic chemicals out, synthesis of detoxification enzymes) or in structural costs (altered metabolic pathways, excretory structures etc.).

Alleles determining insecticide resistance have often been associated with deleterious fitness effects when the insecticide is absent. Roush and Plapp (1982) investigated development time and fecundity in house fly strains that were resistant or susceptible to organophosphates. The fitness of homozygous R strains was reduced by 11 to 43 per cent, but the fitness of the heterozygous genotype was similar to that of the susceptible homozygote. Similarly, Emeka-Ejiofor *et al.* (1983) examined the effects of a semi-dominant dieldrin resistance gene on larval development time in mosquitoes: resistant homozygotes developed more slowly than susceptible homozygotes. In general, deleterious fitness effects of resistance alleles are fairly small and the heterozygous genotype tends to have an equivalent fitness to the susceptible homozygote (Roush and McKenzie 1987), although fitness tests under field conditions have not been carried out and the low frequency of resistance alleles in unsprayed populations suggests that these alleles are generally deleterious. The physiological basis of the deleterious effects is not known, although relatively large deleterious effects may occur when resistance is mediated via esterases, which often constitute a sizable fraction of cellular protein (Roush and McKenzie 1987).

Deleterious effects have been associated with genes providing resistance to chemicals for the control of other pests. Holt and Radosevich (1983) compared the growth rate of two groundsel (*Senecio vulgaris*) biotypes resistant or susceptible to a triazine herbicide, and found that susceptible plants showed a higher initial growth rate, as well as an increase in dry matter production, height and leaf area. Rapeseed (*Brassica napus*) cultivars that are resistant to a triazine herbicide yield 20 to 30 per cent less than susceptible cultivars (Beversdorf *et al.* 1988).

In a study of *Rattus norvegicus* (Partridge 1979) frequency of resistance to warfarin in a large polymorphic population fell from 80 to 33 per cent over 18 months (Table 7.2), from which relative fitnesses were estimated to be 0.46 for the resistant homozygote and 0.77 for the heterozygote. This large deleterious effect for the resistance allele may be related to an increased vitamin K requirement. Resistance to fungicides may also be associated with large deleterious effects: Fuchs *et al.* (1977) examined the resistance of 50 strains of *Cladosporium cucumerium* to triforine, and found a negative correlation between fungicide resistance and growth rate in the absence of the fungicide. In addition, spores from many of the resistant strains were inviable. The authors suggested that the low survival rate of resistant strains may account for the absence of the development of resistance to triforine under natural conditions.

Table 7.2. Proportion of rats resistant to warfarin in samples from the Upper Min-y-llin farm buildings. (After Partridge 1979.)

Month	Susceptible	Resistant	
		No.	%
November 1976	0	12	100
February 1977	8	22	73.3
May 1977	20	45	69.2
August 1977	25	26	51.0
December 1977	13	12	48.0
April 1978	22	11	33.3

Several studies indicate that plants resistant to heavy metal carry a cost in uncontaminated environments. Hickey and McNeilly (1975) compared the performance of normal and resistant plants of four species (*Agrostis tenuis, Anthoxanthum odoratum, Plantago lanceolata* and *Rumex acetosa*) in competition with ryegrass in the field, measuring fitness as dry weight. Resistant plants had extremely low relative fitness values, ranging from 0.32 to 0.001. A limitation of this experiment was that resistant and normal plants came from different populations, and replicate populations were not sampled. In contrast, Wilson (1988) examined resistance in the grass, *Agrostis capillaris*, by collecting two tillers from eight sites representing heavy metal and normal soil types. Resistance of subclones was measured by root elongation in a solution with copper or lead. Growth rate was measured in optimal conditions in pots with high fertility compost. The regression of growth rate onto copper resistance was significant for genotypes (tillers) and populations

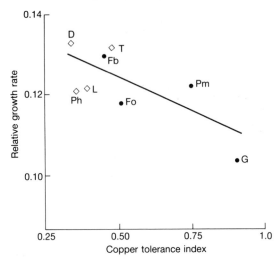

Fig. 7.1. Relationship between copper resistance and relative growth rate in eight populations of *Agrostis capillaris* from mine populations (filled circles) or normal soils (open circles). The regression line is given. (Simplified from Wilson 1988.)

(Fig. 7.1), while the regression for lead tolerance was only significant at the genotype level. Regression slopes were negative in each case, indicating that resistant plants had slower growth rates.

It is not clear whether the poorer performance of plant populations resistant to heavy metals is due to resistance genes or to other genes selected for on mine tailings, such as low levels of nutrients, which may impose an additional selection pressure. Evidence from *Agrostis capillaris* suggests that resistance genes themselves may not be associated with large deleterious effects: copper resistant individuals selected from a cultivar did not differ in dry matter yield from individuals with low resistance (Nicholls and McNeilly 1985). In addition, genotypes of the moss, *Funaria hygrometrica* that were resistant to copper and zinc had a growth advantage, regardless of whether plants were cultured in medium with or without the heavy metals (Shaw 1988).

Deleterious effects associated with heavy metal resistance are not confined to plants. Weis and Weis (1989) described a population of mummichogs or killifish (*Fundulus heteroclitus*) that had evolved increased embryonic resistance to mercury. The increase in resistance was associated with a decrease in adult growth rate and longevity, as well as a reduced ability to capture prey.

There is less information on resistance to chemical stresses not related to human activities. Low concentrations of salt provide one stress that is commonly encountered by marine organisms in estuarine habitats. In the mussel, *Mytilus edulis*, an allele at the *Lap* locus counters the high salinities found in marine environments but causes wasteful nitrogen excretion at low salinity levels, resulting in a reduced growth rate and increased mortality (Hilbish and Koehn 1985; Section 3.3). This may account for clines in the frequency of this allele with changes in salinity.

In summary, many of the above studies indicate that genotypes with increased resistance to chemical stresses will perform less well than susceptible genotypes in the absence of the stressor. The studies with heavy metals and insecticides suggest that fitness differences may not always be large, although these differences have usually not been measured under field conditions. The underlying physiological basis for costs associated with increased resistance is generally not known.

7.3 Genotype–environment interactions

There are two ways of examining trade-offs between stress resistance and other traits when specific genotypes controlling resistance have not been identified. Firstly, we can consider trade-offs between two or more environments when the same trait is used to measure fitness in the different environments. Many of the measures of stress resistance that we have discussed in previous chapters are related to longevity, survival or reproductive output, and these measures can be assessed in optimal as well as in stressful environments. The second way is to look at trade-offs between measures of stress resistance and other fitness-related traits, emphasizing the association between traits rather than environments.

Changes in the performance of genotypes under more than one environment are usually examined with genotype–environment interactions, which test whether differences between genotypes depend on the environment. For trade-offs to exist, these interactions must take the form of a 'crossing type'; that is, the ranking of genotypes changes between environments. Fig. 7.2 considers the performance of two genotypes under a range of environmental conditions where the interaction is of a crossing type. Genotype B is more resistant to stress than genotype A, but has inferior performance under optimal conditions. B performs better than A in region 2, while the reverse is true in region 1. This means that a switch in genotype performances will occur for any comparison of environments that spans regions 1 and 2. We can translate this diagram into a genetic correlation between environments (Falconer 1952) by plotting performance (Fig. 7.3a). If only the two genotypes segregate in a population then a negative genetic correlation between the two environments would be evident because genotypes that perform better in an optimal environment perform more poorly in a stressful one. A similar situation where more than two genotypes are involved is

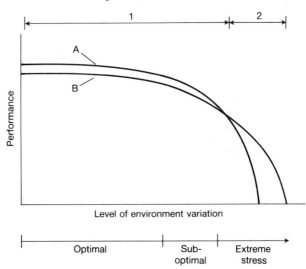

Fig. 7.2. Performance of two genotypes at different levels of an environmental variable which is stressful at high levels. Genotype A has higher relative performance under optimal conditions while genotype B is more stress resistant.

illustrated in Fig. 7.3b. The same approach can be used to visualize a trade-off between different traits. Note that positive genetic correlations will occur for much of the environmental region in Fig. 7.2; negative correlations will only be evident when environmental conditions become sufficiently extreme. Although the sizes of regions 1 and 2 are not known exactly, those given in Fig. 7.2 are appropriate when organisms can function over a wide range of environmental conditions and differences between genotypes for stress resistance are only evident at extremes, as suggested by some of the data discussed in Chapter 5.

Trade-offs can only be determined from genotype–environment interactions when traits are considered in the context of fitness. Demonstrating that the same genotypes have relatively high scores on a trait over a range of environmental conditions does not rule out the possibility of a trade-off. If traits other than fitness components are used, these must be related to fitness under the conditions where they are measured. For example, Huey and Hertz (1984) considered locomotion in lizards at a range of body temperatures, including the temperature range likely to be experienced in nature, and found that the same individuals tended to have high or low running performance at all temperatures. Although the authors conclude that there is no trade-off at the phenotypic level, it is possible that running performance may not be positively associated with fitness at all temperatures. Running increases metabolic rate and may therefore incur unnecessary energy expenditure

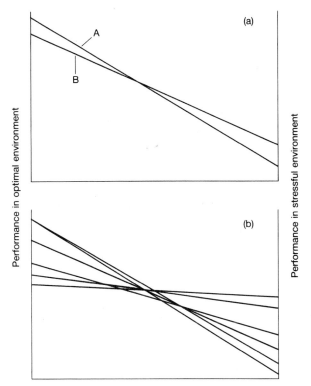

Fig. 7.3. (a) Performance of two genotypes under optimal and stressful conditions given by regions 1 and 2 in Fig. 7.2. (b) Trade-off between stress resistance and performance under optimal conditions as indicated by scores from six genotypes.

under conditions when survival depends on a reduction in energy expenditure during hot periods. Physiological performance needs to be associated with fitness under a range of natural conditions, and such associations have only been demonstrated in some cases (Pough 1989).

A few studies have considered genetic variation for fitness-related traits under a range of environmental conditions that encompass stressful as well as optimal situations. The most extensive data on genotype–environment interactions come from the agricultural literature, particularly in crop plants. Unfortunately, agricultural studies are usually limited to comparisons of a few select varieties rather than a random sample of genotypes from a population, and conditions of 'environmental stress' are usually defined loosely as those that do not result in an optimal yield, rather than the conditions close to lethality discussed in Section 5.5. These limitations need to be kept in mind when extrapolating from agricultural studies to natural populations.

7.3.1 *Agricultural studies: plants*

Analysis of genotype–environment interactions in plant breeding experiments has often been based on the regression technique of Finlay and Wilkinson (1963), which uses the mean yield of cultivars at a site as a measure of the environment; performance of each cultivar is regressed against this measure after appropriate transformation of the data. Stressful environments are defined as those that show large departures in mean yield from the highest yielding environment. Eberhardt and Russell (1966) extended this analysis by suggesting that deviations from the regression were also important because they provided a measure of a cultivar's stability of production across environments. Both approaches were used to identify ideal cultivars in plant breeding programmes. Finlay and Wilkinson (1963) proposed that an ideal widely adapted cultivar should have a small regression slope (b) and maximum yield in the most favourable environment: this would ensure that cultivars would still have a reasonable performance under low yielding conditions. In contrast, Keim and Kronstad (1979) emphasized performance under stressful conditions and suggested that an ideal cultivar should have the highest yield in the most stressful environment and a high b value, reflecting a strong response to favorable conditions. If trade-offs exist between stressful and optimal environments then these ideals cannot both be attained, at least under conditions when the yield suppression (environmental stress) becomes severe. Performance in stressful or optimal environments and the regression slope are not expected to evolve independently in the presence of trade-offs. Strains that have high performance under favourable conditions should have high b values, while strains that show high performance under stressful conditions should have low b values, as long as the range of environments is sufficiently broad to encompass extreme conditions.

Some studies with cultivars suggest that mean performance, regression response and performance stability are genetically independent characters or may even be positively correlated, implying that trade-offs between environments do not occur. Langer *et al.* (1979) found that these characters were independent in a sample of oat varieties grown under a range of environments. Bruckner and Frohberg (1987) considered yield performance in 20 cultivars of spring wheat grown under undefined environments that provided yields ranging from 4815 to 285 kg/ha and found a range of responses to environmental conditions and no indication of a trade-off between performance under stressful and non-stressful conditions. However other studies suggest that the 'ideal' cultivar, with high performance in stressful and favourable conditions, may not exist. Keim and Kronstad (1979) characterized nine wheat cultivars exposed to a wide range of moisture-limited conditions where mean yield varied from 940 kg/ha to 4820 kg/ha. Varieties with high performance under stressful conditions tended to have low regression slopes as predicted by trade-offs between environments. Fischer and Wood (1979) considered 34 wheat cultivars under drought-stressed conditions that reduced yield by 60 per cent. Cultivars which showed the greatest decrease under drought had relatively higher non-drought yields, as reflected by a genetic correlation of 0.76 between these traits.

Since cultivars do not represent a random sample of genotypes from populations it may be difficult to interpret findings in terms of genetic correlations in populations. In addition, most cultivars have been developed for their performance under optimal conditions, and there has not been much attempt to breed for increased performance under stress (Buddenhagen 1983; Foy 1984). Directional selection for yield under optimal conditions will have eliminated many genes that increase yield under stress, and decrease yield under optimal conditions. The range of environmental conditions in many experiments may not be broad enough to encompass stressful and optimal situations, and Ceccarelli (1989) suggested that experiments encompassing environments with extremely low yields are more likely to provide evidence that different genotypes are favoured under stressful conditions.

Selection experiments may provide a better indication of trade-offs than comparisons of cultivars because they involve individuals from the same population. Selection is carried out in a stressful and optimal environment and the performance of selected lines is compared in a range of environments. Ceccarelli (1989) reviewed selection experiments for yield under extreme conditions, and concluded that different genotypes tended to be selected under stressful and optimal conditions, with the possibility of trade-offs in some cases. For example, Ceccarelli (1987) found evidence for a trade-off between environments in barley. When families were selected for grain yield in favourable and unfavourable (dry) conditions, the average yield was 873 kg/ha at the driest site and 3959 kg/ha at the wettest site. Comparison of yield and drought susceptibility (decrease in yield compared to yield without drought) for the selected groups at the dry and wet sites showed a strong negative correlation ($r = -0.78$) between yield and drought susceptibility at the driest site, indicating that yield reflected drought resistance. In contrast, this correlation was smaller and positive at the wetter site ($r = 0.25$), indicating that high yield under favourable conditions was associated with decreased drought resistance. Genes increasing yield in favourable conditions therefore decreased yield under stressful conditions.

Results from the Park Grass experiment, where a population of *Anthoxanthum odoratum* has been maintained for 60 years in a patchy environment (Section 4.4), also suggest trade-offs, although these have arisen as a consequence of natural rather than artificial selection. Davies and Snaydon (1976) carried out reciprocal transplants between plants from adjacent pairs of plots that differed in fertilizer treatments and liming, which have a large influence on yield and vegetation height (Table 7.3). Plants tended to survive longer, produce more tillers, and produce more dry matter on their native plots than on the contrasting plots, despite the overall yield differences between plots (Fig. 7.4), suggesting the evolution of trade-offs in performance between environments.

A possible trade-off between yields in optimal and in stressful environments is important in future agricultural developments because past selection for high yielding varieties under conditions where environmental fluctuations are minimized may have resulted in plants that expend the maximum amount of energy possible on reproductive effort (in the case of grain crops) and minimal effort on maintenance functions or structures associated with stress resistance. Hanson and Nelsen (1980) illustrate this problem with respect to dwarf wheat and rice varieties developed during the green revolution, which were bred for short stature to prevent lodging under optimal conditions. The shortening and reduction in lower leaves increased yield, probably because of decreased allocation to vegetative structures, but made the plant more prone to competition from weeds. The less extensive root system of these plants also made them more susceptible to water stress. Green revolution cultivars also tend to perform poorly on nutrient-deficient soils (Foy 1984).

7.3.2 Agricultural and laboratory studies: animals

Experiments with animals have provided ambiguous results on the importance of genotype–environment interactions (Pani and Lasley

Table 7.3. Environmental conditions on source plots of three reciprocal pairs of populations of *Anthoxanthum odoratum* from the Park Grass experiment. (Simplified from Davies and Snaydon 1976.)

Population pairs	Fertilizer treatment	Mean annual yield (kg/ha)	Vegetation height (mm)	Soil pH
9 (limed)	N, P, K, Na, Mg	5613	580	5.3
3 (unlimed)	Unfertilized	1475	110	5.2
1 (limed)	N	2363	160	7.2
1 (unlimed)	N	1700	120	4.0
8 (limed)	P, Na, Mg	2150	230	7.0
7 (unlimed)	P, K, Na, Mg	3650	270	4.9

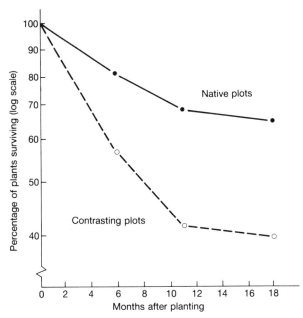

Fig. 7.4. Mean survival of *Anthoxanthum odoratum* from the Park Grass experiment transplanted into their native plots (filled circles) and into contrasting plots (open circles). (Adapted from Davies and Snaydon 1976.)

1972). Some selection experiments provide evidence for large interactions: Falconer (1960) selected for growth between 3 and 6 weeks of age of mice on a high or low level of nutrition. Animals on the low level of nutrition were given food cubes diluted with 50 per cent of indigestible fibre: their growth rate was reduced by about 20 per cent. Growth at the high nutrient level was increased equally by selection at the low and high levels, but growth at the low level was only increased by selection at this level. Growth increases were reached by different physiological pathways, suggesting a different genetic basis to these responses. Hardin and Bell (1967) selected for body weight in *Tribolium* maintained at two levels of nutrition. Animals kept at the high level of nutrition attained a body size twice that attained at the low nutrient level. The genetic correlation between the environments was 0.60 ± 0.21, indicating that much of the selection response involved genes whose contribution to increased body weight was specific to one level of nutrition.

Other selection experiments have found no evidence for genotype–environment interactions. Bohren *et al.* (1981) selected for fast and slow growth in poultry at a non-stressful temperature (21.1°C) and a hot temperature (32.2°C): the latter produced a 20 per cent reduction in weight gain. There was no genotype–environment interaction when lines

were tested in different environments; selection in one environment led
to a similar selection response in the other environment. Van Vleck
(1963) considered milk yield in dairy cattle sired artificially over a
9-year period, and classified seasons into four levels depending on the
overall milk yield. Yields at the most extreme levels varied by about 20
per cent. Estimates of genetic correlations between environments varied
between 0.94 and 1.04, indicating that genotype–environment interactions for milk yield were generally absent.

None of these experiments provide evidence for trade-offs between
environments because negative genetic correlations have not been
demonstrated, even in the experiments where genotype–environment
interactions existed. However these experiments provide only a limited
test of trade-offs between environments. One problem with the
experimental design of many of the selection experiments is that selection lines were unreplicated: when only one line was selected in each
environment the effects of genetic drift cannot be evaluated (Section
6.5). Another limitation is that some of the performance traits, particularly those measured in agricultural studies, may not be positively correlated with fitness in all environments. High rates of reproduction may
lead to decreased survival under stressful conditions because of the
energy allocated to reproduction, and individuals with lower short-term
reproductive outputs may have higher lifetime outputs under stressful
conditions. This emphasizes the need to express trade-offs between
environments in terms of the contribution that traits make to fitness in
each environment.

A third limitation is that many experiments use non-optimal environments which are not particularly stressful. A 20 per cent reduction in
growth does not place an organism under a severe stress close to
lethality. Such yield reductions may be important from the agricultural
perspective that provided the motivation for many of the experiments,
but they may not reflect the difference between optimal and stressful
environments encountered in nature. If differences in the performance
of genotypes follow the type of curve illustrated in Fig. 7.1, trade-offs
will only be evident when experimental conditions encompass extremes.
For example, growth rate in cattle shows only small genotype–environment interactions when breeds are compared within temperate environments, but rankings of breeds for growth rate can change when
comparisons are made across temperate and tropical regions (Warwick
1972).

A trade-off for growth rate in cattle between stressful and non-
stressful conditions was demonstrated in a selection experiment (Frisch
1981) mentioned in Section 6.2. Lines selected for high weight gain
under stressful field conditions (high temperature, high parasite load,

low nutrition) were tested under conditions of low stress: in pens with food *ad libitum*, and after dipping and drenching to remove parasites. Weight gain, measured in winter and spring when temperatures were not stressful, was significantly lower for steers from the selected line than for steers from the control line under these optimal conditions. The selected line had a lower mass-specific metabolic rate under fasting than the control line. This reduction in metabolic rate may form the basis for the trade-off between environments because growth rate of cattle breeds under non-stressful conditions is positively correlated with their fasting metabolic rate (Frisch and Vercoe 1978).

More genotypic comparisons across a range of stress levels are needed, where stressful conditions can be demonstrated on the basis of mortality, reduced adenylate energy charge, or some other physiological indicator. The term 'stress' has been applied too often to slight changes in environmental conditions unlikely to affect the adenylate energy charge (Section 1.1). The generality of genetic trade-offs between environments will not be clear without further research, although some data and physiological considerations discussed above suggest that trade-offs may occur when stress levels are extreme.

7.3.3 Data from natural populations

Genetic data from within natural plant and animal populations are scarce, although some indirect evidence suggests trade-offs between stressful and optimal environments. One line of evidence comes from genetically based temporal changes in the stress resistance of a population. Stress-resistant genotypes are at a disadvantage at some times of the year if resistance increases in some seasons and decreases in others. A few examples of seasonal changes in gene frequencies were reviewed in Section 4.4, Carvalho's (1987) study of *Daphnia* being a particularly convincing case. Other temporal changes suggesting a trade-off include rapid decreases in the frequency of stress-resistant genotypes once a stress is removed, as in the rapid decrease in warfarin resistance in rats, discussed earlier in this chapter.

Microspatial genetic variation in stress resistance provides further indirect evidence for trade-offs. If stress-resistant genotypes are favoured in one habitat and there is gene flow between this habitat and more favourable habitats, then spatial differentiation should only persist if resistant genotypes are selected against in the favourable habitats. Resistant genotypes would otherwise generally increase in frequency. We discussed evidence for microspatial variation in Chapters 4 and 6, including responses to heavy metals and gradients in moisture and climatic factors which suggested that this type of variation is common in natural populations.

Many geographic comparisons of plant and animal populations suggest that populations perform better in the environment from which they originated (Section 4.2) and these differences are often considered to reflect trade-offs. For example, Browne *et al.* (1988) measured 14 life-history traits at three temperatures in populations of brine shrimp from different geographic locations. The rankings of populations often switched at the different temperatures, and no single population had superior performance across all temperatures. Baldwin and Dingle (1986) compared life-history traits in two populations of the milkweed bug, *Oncopeltus fasciatus*, from different thermal regimens. The population from the cooler environment showed increased survival and reproduction at a cool temperature, and performed more poorly for some traits at warmer temperatures. Both studies were interpreted as indicating that populations performing better in one environment performed more poorly in another. However, such interpretations should be made cautiously because there are situations where apparent trade-offs at the geographic level will not reflect negative genetic correlations. Populations may diverge when there is no trade-off because one set of genes is favoured in one environment and a different set is favoured in another environment. Populations will then perform better in their own environment once they have diverged. Genetic differences between populations may also be a consequence of genetic drift or mutational change. 'Trade-offs' at the geographic level may therefore reflect a different history of selection and random processes causing population divergence as well as genetic trade-offs.

The most direct way of testing genetic trade-offs in natural populations is to follow parental and offspring generations in the field. Parent–offspring correlations can be examined when the offspring experience different conditions from those experienced by the parents. Low heritabilities would indicate that different genes are responsible for high or low expression of a trait in conditions experienced by parents or offspring, while negative heritabilities would suggest trade-offs. Another approach is to transfer members of the same sib groups to stressful and favourable field situations. The feasibility of obtaining field estimates for some organisms is indicated by field studies on body size in natural populations of the great tit (*Parus major*), where heritability was shown to depend on environmental conditions (van Noordwijk *et al.* 1988). Heritability estimates based on the correlation between parent size and size of offspring ranged from 0 to around 0.5. The lower values corresponded to growth of nestlings under poor feeding conditions when considerable mortality (≥ 25 per cent) occurred, suggesting that stress decreased the heritability or increased the size of genotype–environment interactions.

Finally, genetic trade-offs between environments may be indicated by the performance of specific genotypes under different conditions. The data discussed in Section 3.1 indicate that an enzyme morph which performs better than an alternative morph at a given temperature is usually inferior to it at another temperature, and similar findings have been made for genetic polymorphisms within populations. For example, the α-haemoglobin chains of the deer mouse, *Peromyscus maniculatus*, are encoded by two linked loci, and two haplotypes predominate, a_0c_0, which is common at high altitudes and a_1c_1, which predominates at low altitudes (Chappell and Snyder 1984). Mice with the a_1c_1/a_1c_1 genotype have higher oxygen consumption rates at low altitudes, while the a_0c_0/a_0c_0 homozygote has higher oxygen consumption rates at high altitudes; heterozygotes are intermediate. Increased oxygen consumption is probably related to increased resistance to hypoxia and cold conditions experienced by deer mice at high altitudes. However the extent to which specific genotypes contribute to the phenotypic variance of a fitness trait is not usually determined in such studies so that the contribution of particular genotypes to genetic correlations is not known.

7.4 Trade-offs between traits

We now turn to trade-offs between stress resistance and other traits related to fitness under favourable conditions which, as mentioned earlier, may reflect resource allocation patterns or the conservation of resources. Because resource conservation is associated with a reduced metabolic rate, we will start by briefly considering some of the associations that have been demonstrated between metabolic rate and fitness-related traits.

In mammals, there has been much discussion of a positive association between population growth rate as measured by the Malthusian parameter, r_m, and the resting metabolic rate (McNab 1980; Fig. 7.5). The parameter r_m is defined as

$$r_m = (\ln R_0)/T,$$

where R_0 is the maximal rate of reproduction per generation and T is the minimum generation time; these in turn reflect postnatal growth, number of offspring and gestation time. Generation time varies inversely with metabolic rate independently of mass in mammals, while the product of litter size and number of litters and growth rate varies positively with resting metabolic rate (McNab 1980). In mammalian species, the rate of basal metabolism appears to be the highest that can be sustained by the

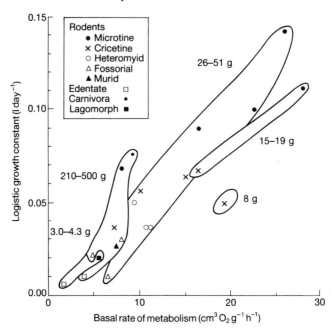

Fig. 7.5. Association between the postnatal logistic growth constant and the basal rate of metabolism in mammals within ranges of body size. (After McNab 1980.)

quantity and quality of their food resources, to maximize their reproduction. Insectivorous bats and desert mammals that have variable food supplies show decreased metabolic rates compared to related species with less variable food sources (McNab 1979). A high metabolic efficiency, indicated by a low resting metabolic rate is not favoured when resources are abundant. Instead, resting metabolic rate is positively correlated with maximal or mean daily metabolic rates, so that it can be used as an indicator of total energy expenditure (McNab 1980); high rates of energy expenditure are associated with high rates of increase. The maximization of growth rate at the expense of efficiency is also evident in situations where organisms change from efficient metabolism to a high rate of metabolism at different ontogenetic stages. McClure and Randolph (1980) showed that the rodent, *Sigmodon hispidus*, became homeothermic at a smaller size and earlier age than a similar species (*Neotoma floridana*), resulting in a lower growth efficiency but a higher growth rate. This was related to differences in food supply: *S. hispidus* has a continuous food supply, whereas *N. floridana* experiences periods of food shortage.

There are few interspecific comparisons of the association between growth rate and metabolic rate in ectotherms, although some data tend to support the mammalian findings. Branch *et al.* (1988) found that oxygen consumption of intertidal organisms is related to food availability (Section 6.1) and is positively associated with growth. For example, the limpet species, *Patella cochlear*, has a low rate of oxygen consumption and grows 3 mm per year, while the related species, *P. oculus*, has a high rate of oxygen consumption and grows 30–60 mm per year. A positive association between growth rate, reproductive rate and metabolic rate has also been described in lizards (Anderson and Karasov 1981; Nagy 1982).

Physiological correlates of growth rate have been considered at the intraspecific level. Several genetic experiments bearing on this question and motivated by agricultural applications were carried out in mice. Variation in growth rate in mice may occur through changes in feeding rate, capacity for protein deposition, allocation of energy to growth, and efficiency of energy conversion (McCarthy 1980), factors which are not independent and which may not relate to metabolic rate in the relatively straightforward manner discussed above. For example, an increased feeding rate will increase the active metabolic rate, and may decrease the efficiency of energy conversion if the digestive system becomes overloaded. However an increased feeding rate may also increase fat deposition at the expense of protein deposition, and this may in turn influence active and resting metabolic rates because of the variable maintenance requirements of tissues with a different composition. In general, selection experiments for increased growth rate or body weight have resulted in increased feeding rate (and hence an increase in the active metabolic rate), but changes in other factors are less consistent (McCarthy 1980; Malik 1984). Fat deposition rather than protein deposition has been shown to increase in several selected lines. Some experiments suggest decreased maintenance requirements, but these may often be associated with changes in body weight rather than changes in weight-specific metabolic rate.

Rapid rates of metabolism should increase energy production and anabolic processes which increase reproductive output, and decrease the time organisms take to reach maturity. A few studies have directly related genetic variation in the rate of metabolic flux through biochemical pathways to variation in fitness. In chemostat experiments with *E. coli* under conditions of competition for lactose (Dykhuizen *et al.* 1987), growth rate is proportional to the rate of lactose metabolism, favouring strains that constitutively produce enzymes involved in this process and strains with high activities of these enzymes. In *Colias* butterflies, genotypes at the phosphoglucose isomerase locus that increase flux are favoured by selection (Watt 1985). Genetic differences in rates of flight metabolism between isogenic chromosome lines of *D. melanogaster* show a correlation of >0.9 with the mechanical power output of flight muscles, a measure of flight ability (Laurie-Ahlberg *et al.* 1985) which is related to reproductive success because it influences the ability of flies to locate food and mates. Decreased development time (and hence increased fitness) is likely to be associated with increased flux when food is available *ad libitum*, because increased activity of many enzymes is generally correlated with preadult development rate in *D. melanogaster* and *D. subobscura* (Marinkovic and Ayala 1986; Marinkovic *et al.* 1987; Cluster *et al.* 1987). As a final example, hatching time in the fish *Fundulus heteroclitus* depends on variation at the lactase dehydrogenase locus; the genotype with the fastest development time has the highest metabolic rate (DiMichele and Powers 1984).

Metabolic rate may be associated with fitness-related traits other than growth and reproductive output. Evidence from the behavioural literature indicates that social status and territorial defence are positively associated with metabolic rate. The weight-specific resting metabolic rates in winter flocks of great tits (*Parus major*) and in pied flycatchers (*Ficedula hypoleuca*) on breeding territories were associated with dominance status (Fig. 7.6; Roskaft *et al.* 1986), suggesting that dominant birds have a higher energy demand. Metabolic rate has also

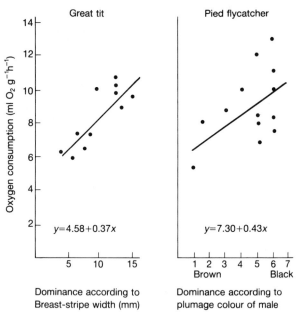

Fig. 7.6. Relationship between social dominance and oxygen consumption in the great tit and the pied flycatcher. (Adapted from Roskaft *et al.* 1986.)

been correlated with dominance status in willow tits, whereby it changed according to the dominance rankings of the birds (Hogstad 1987). In the fish, *Betta splendens*, which forms hierarchies in tanks, winners and dominant individuals produced more energy (increased glycogen, protein, and free glucose content, increased carbohydrate degradation) than losers and submissive individuals (Haller and Wittenberger 1988). None of these studies have considered variation at the genetic level and more work is needed to determine whether social status reflects differences in metabolic rate or if these differences are a consequence of social status.

Metabolic rate has also been considered in research on longevity, particularly in aging studies in houseflies, *Drosophila* and mice. In general, insects that live longer have lower metabolic rates, and conditions that increase metabolic rate tend to reduce lifespan (Sohal 1986; Sohal *et al.* 1987). Sacher and Duffy (1979) examined the association between longevity, body weight and oxygen consumption in 21 of the possible 25 matings between five inbred strains of mice. Lifespan was negatively correlated with average metabolic rate (over a 24 hour

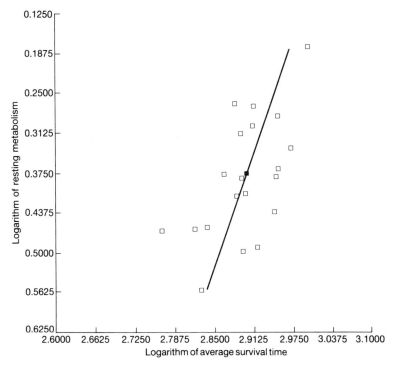

Fig. 7.7. Relationship between mean survival time and resting metabolic rate in strains of mice. (From Sacher and Duffy 1979.)

period) and resting metabolic rate, and positively correlated with body weight, regardless of whether measurements were made in young or old mice (Fig. 7.7). One explanation for these results is that longevity is affected by oxidative stress; a lower metabolic rate is one of the factors that reduces this stress.

Finally, metabolic rate may be associated with competitive ability. Mueller (1988) successfully selected for competitive ability for food in *D. melanogaster* larvae. The outcome of larval competition was determined by larval feeding rates. Joshi and Mueller (1988) found that larvae from strains selected for increased competitive ability had higher feeding rates than larvae selected under an abundant food supply. Larvae that consumed the most food were more likely to reach pupation. The increased feeding activity would be associated with an increased active metabolic rate, and Mueller suggested that this may result in decreased metabolic efficiency, although measurements of active and resting metabolic rates were not made.

These results indicate that energy conservation and resource allocation could provide a basis for trade-offs between fitness-related traits and measures of stress resistance. There is much non-genetic evidence for an interaction between stress resistance and performance on life history traits. The interspecific association between slow growth rate and increased stress resistance in plants and some animals (Section 6.3) indicates that resistant organisms will often be out-competed under favourable conditions. Comparisons of different life cycle stages generally indicate that fast growth reduces stress resistance because rapidly growing juvenile stages or reproductive stages tend to be more susceptible to stress damage than slowly growing stages. Experimental manipulations often indicate an association between stress resistance and life history traits, in that stressful treatments at one life cycle stage often decrease subsequent performance on life-history traits under optimal conditions. For example, exposure to temperature extremes at the pre-adult stage in *D. melanogaster* reduces adult longevity, female fecundity and male fertility at the optimal temperature of 25°C (David *et al.* 1983). Conversely, enhanced reproductive activity can decrease stress resistance, as shown by the lower starvation resistance of *D. melanogaster* males held with females compared to virgin males (Service 1989).

Unfortunately, most studies on trade-offs at the genetic level have examined associations between reproduction and longevity or growth and reproduction rather than associations between stress resistance and these traits. Stress resistance and longevity were associated in the hymenopteran parasite, *Aphytis lingnanensis* (White *et al.* 1970). Lines selected for increased resistance to temperature extremes lived 1.5 to

2.5 times as long as control lines; this difference was evident over a range of temperatures. There is also some evidence for genetic correlations between stress resistance, adult longevity and early female fecundity in *Drosophila melanogaster*. Lines selected for postponed senescence show a decrease in early fecundity, suggesting a genetic trade-off between these life-history traits (Rose 1984; Luckinbill *et al.* 1984). The lines selected by Rose were characterized for stress resistance by Service *et al.* (1985) and Service (1987), and found to have increased resistance to starvation, desiccation and a toxic concentration of ethanol (Chapter 6). Reverse selection for early-age fecundity led to a correlated response for decreased starvation resistance (Service *et al.* 1988), suggesting that the same genes affected resistance to starvation, longevity and early fecundity. The *D. melanogaster* lines that we selected for increased stress resistance also showed changes in early fecundity, the selected lines having a lower fecundity than control lines (Hoffmann and Parsons 1989a). These correlations between traits may be related to changes in resting metabolic rate because there was a reduction in metabolic rate in our lines and in the lines characterized by Service (1987). However, Arking *et al.* (1988) found no differences in metabolic rate between a different set of long-lived lines and short-lived lines over the lifetime of the flies, and genetic variation in longevity in *D. melanogaster* may be associated with other factors such as lipid and glycogen synthesis (Service 1987; Luckinbill *et al.* 1988, 1989).

Coat colour and coat type in cattle have high heritabilities and provide the basis for a trade-off between resistance to heat stress and growth rate. Individuals with light coat colours are more resistant to heat stress than dark individuals (Section 4.1), but cattle with dark coats have relatively higher growth rates under good grazing conditions and in cold regions (Finch and Western 1977). Black cattle may save metabolic energy by absorbing radiant energy, thereby reducing the energy expended on cold thermogenesis. Cattle with sleek coats have higher growth rates in the tropics and a decreased susceptibility to heat stress compared to cattle with more woolly coats, but the latter are considered the best producers under cooler, more favourable conditions (Turner and Schleger 1960).

An association between heat resistance and decreased size in endotherms has been demonstrated in selection experiments for increased heat resistance in mice (Jonsson *et al.* 1988) and chickens (Wilson *et al.* 1975). Decreased size reflects a decreased growth rate, which will lead to a reduction in reproductive output under optimal conditions.

Most of the genetic experiments with plants have considered trade-offs between environments rather than between traits. Nevertheless, the studies discussed in Section 6.3 suggest a negative association between stress resistance and slow growth rate at the genetic level, such as the association between growth rate and survival at low temperatures in tall fescue (*Festuca arundinacea*) varieties (Robson and Jewiss 1968). Slow growth should lead to a reduced competitive ability under optimal conditions (Table 7.1), indicating the potential for genetic trade offs with other fitness-related components.

The above discussion suggests the likelihood of genetic trade-offs between stress resistance and other fitness-related traits measured under favourable conditions. Some of the trade-offs may reflect the positive association between metabolic rate and fitness under optimal conditions, but additional genetic studies are required, and this prediction may not hold when resistance is associated with metabolic efficiency as discussed in the next section.

7.5 Heterozygosity, resistance, and metabolic efficiency

The possibility of an association between stress resistance and increased metabolic efficiency (Section 7.1) suggests that increased resistance may not always show genetic trade-offs with fitness measured under favourable conditions. Koehn and Bayne (1989) suggested that the association between heterozygosity and growth rate may underly a positive genetic correlation between stress resistance and growth rate. They provided a model where a genotype with a lower maintenance requirement can support growth over a wider range of environmental conditions. This model was based on studies of heterozygosity at enzyme loci in organisms such as the mussel, *Mytilus edulis* (Koehn and Shumway 1982). Heterozygous mussels appear to have decreased energy requirements for maintenance because of greater efficiency in protein synthesis. The conserved energy is available to increase effort for ingestion and absorption (Hawkins *et al.* 1986). This is consistent with evidence that loci having large and significant correlations with growth rate in the coot clam (*Mulinia lateralis*) code for enzymes which function in protein catabolism or glycolysis, and not for enzymes which function in the pentose shunt or in maintaining redox balance (Koehn *et al.* 1988). The growth advantage of highly heterozygous individuals may be enhanced by heterogeneous environments and under stressful conditions where the conservation of energy becomes particularly important. This has already been discussed (Section 5.3), and is illustrated by the conservation of weight by more heterozygous mussels under starvation (Diehl *et al.* 1986).

Supporting evidence for a positive correlation between heterozygosity and growth has been found in marine invertebrates, the tiger salamander, white-tailed deer, pigs, sheep, man, aspen and several species of conifer (Mitton and Grant 1984), although there are some exceptions to this rule and the association is weak in several cases. The relationship between heterozygosity and maintenance oxygen consumption has only been examined in a few cases, and extrapolations from data on marine invertebrates should be made cautiously. One feature that distinguishes *M. edulis* is its relatively high resistance to anoxia, and some populations of *M. edulis* may have been selected in the low oxygen environments to which they are exposed at low tide (Theede *et al.* 1969). This means that low oxygen levels are an environmental stress for some intertidal populations of *M. edulis*, and differences in oxygen consumption may partly reflect stress resistance. Mussels and clams are also sessile, so that differences between active and resting metabolic rate will be smaller than in non-sessile organisms.

Some results suggest that the relationship between heterozygosity and metabolic rate will differ in other organisms. In *D. melanogaster*, multilocus heterozygosity and development time were not correlated when loci associated with glycolysis were sampled (Marinkovic *et al.* 1987). Danzmann and Ferguson (1988) found that rainbow trout with intermediate development times were more likely to be heterozygous than those at either end of the distribution. Heterozygosity in tiger salamanders was negatively correlated with maintenance oxygen consumption but positively correlated with active oxygen consumption, measured by forcing salamanders to move continuously (Mitton *et al.* 1986). This means that heterozygous individuals have a relatively greater scope for metabolism which may lead to a positive association between heterozygosity and total oxygen consumption.

Two other considerations should also be mentioned. First, the association between heterozygosity and fitness traits may not reflect underlying heterozygote advantage, as discussed in Section 5.3. This question should be resolved if more studies like those of Koehn *et al.* (1988) are carried out and the association between heterozygosity and growth rate is assigned to a few loci. Second, genes that increase efficiency are likely to be favoured under stressful and favourable conditions, and they should go on to fixation. Such genes are therefore unlikely to contribute much to genetic correlations between growth and stress resistance. This may explain why genetic variation in metabolic efficiency has been associated with heterozygosity: such variation will persist if heterozygous genotypes are the most efficient. However the genetic correlation between growth and stress resistance will reflect other genes that are not favoured in both optimal and stressful environments, and these genes may largely dictate the nature of the genetic correlations, particularly as much of the genetic variation in stress resistance traits appears to be additive (Chapter 5).

7.6 Trade-offs and stress evasion

As discussed in Section 4.5, stress evasion often involves changes in life-history traits to minimize exposure to stressful periods. Trade-offs for stress evasion traits can be related to aspects of life-history theory concerned with the response of organisms to environmental stress. Two types of trade-offs can be distinguished: those associated with switching between different life cycle stages, and those associated with the costs of producing an evasion mechanism.

The first type of trade-off depends on the occurrence of life-history stages that differ in stress susceptibility. Organisms are expected to minimize the time spent in a more susceptible stage in order to reduce the effects of a stress. This concept is a component of life-history models that predict the timing of reproduction on the basis of adult survival rates relative to juvenile survival rates (e.g. Charnov and Schaffer 1973). Trade-offs between environments may occur when individuals are exposed to different conditions that alter the relative survival rates of two or more lifecycle stages.

A simple case involves trade-offs associated with dormancy or the production of dormant propagules. Most non-dormant individuals will die when an extreme stress occurs; a dormant life cycle stage is therefore favoured when stressful conditions are common. However individuals without a dormant stage can continue to reproduce under continuously favourable conditions. Stress evasion in the form of dormancy will therefore be favoured in some conditions and selected against in others, forming the basis of a genetic trade-off between optimal and stressful environments in a population that is polymorphic for the production of dormant lifecycle stages. In Section 4.5 we discussed some cases where such polymorphisms occur within a population: dormancy has been related to fitness in Venable's (1984) study of two seed types produced by the annual plant, *Heterotheca latifolia*. The dormant seed type is favoured under drought conditions when competition is low, whereas non-dormant seeds have higher fitness under more favourable conditions when competition is more intense.

This type of trade-off is also relevant to other switches between lifecycle stages where the stages differ in stress resistance. Newman's (1988a,b) study of toads provides an example where timing of metamorphosis is an underlying trait which affects body size as well as maturation time (Section 4.5). Individuals that mature late can exploit resources for longer and may have a larger body size than early maturing individuals, but they run the risk of failing to complete development if ponds dry up too quickly. Genetic variation for timing to metamorphosis therefore forms the basis of a trade-off between environments.

Such trade-offs will result in genetic divergence between populations in different environments. An example is the case of development time and body size in the wood frog, *Rana sylvatica* (Berven and Gill 1983). The frogs breed in small temporary ponds, and metamorphosis is followed by a terrestrial juvenile and adult phase. Larval traits from populations in three locations were considered:

(1) a lowland habitat, characterized by a long favourable growing season but long summers;

(2) a tundra habitat, near the northern distribution limit marked by a short growing season and cold temperatures;

(3) a high elevation site, also with a short growing season and cool temperatures.

Selection for development time is expected to be stronger than selection for size in the habitats with shorter growing seasons because frogs need to avoid stressful conditions, especially because cooler temperatures extend development time but increase body size. The reverse is expected in the lowlands, since rapid development reduces size at metamorphosis, resulting in more intense selection for increased body size. These different selection pressures were reflected in population differences in the means for development time and size. In addition, heritability for body size was low in the lowlands and heritability for development time was low in the tundra, consistent with the direction of selection expected for these traits: heritabilities for both traits were high in the mountain population, where a high heritability for development time was not expected.

The second type of trade-off is associated with the cost of producing evasion mechanisms. Dormant seeds will cost more to produce than non-dormant seeds because they carry food reserves to enable survival through stressful conditions, so allocation of resources to dormant seeds will not be favoured in optimal environments. Plants developing from dormant seeds may also show reduced fecundity and survival compared to plants from non-dormant seeds (e.g. Venable 1984). The development of wings and flight muscles in insects helps to evade stressful conditions (Section 4.5), but winged morphs may be at a disadvantage under favourable conditions. The absence of flight muscles may provide space and energy for organ systems involved in reproduction, and this leads to increased fecundity and more rapid reproductive maturity in flightless morphs (Harrison 1980). In Kaitala's (1988) study of the waterstrider, *Gerris thoracicus*, one of two morphs flies throughout the reproductive period and maintains its flight muscles, while the other histolyses its flight muscles. Non-flier females produced 60 to 100 per cent more eggs than fliers when food was abundant and non-fliers survived longer when food was scarce. Some of the data discussed by Vepsalainen (1978) suggest that such trade-offs lead to genetic divergence for winged morphs between waterstrider populations exposed to different habitats. Similar trade-offs have been described by Denno *et al.* (1989) for winged and flightless planthoppers (*Prokelisia diolus*): winged individuals had a shorter lifespan, reproductive delay, and decreased fecundity.

Phenotypes that enable organisms to evade a stress by switching resources may also be subject to fitness costs in the absence of a stress. Selection for increased body size in Darwin's finches (*Geospiza fortis*) under drought conditions was mentioned as an example of genetic variation in this stress evasion mechanism. Selection of body size was examined after a favourable wet period when seed biomass was high and birds bred for 8 months instead of the usual 1 to 3 months (Gibbs and Grant 1987). Mortality in the subsequent 2 years was higher for larger birds, and this may have been related to the abundance of small seeds, which tend to be foraged by the smaller birds. These results suggest oscillating selection for body size in Darwin's finches mediated by climatic changes because the phenotype involved in stress evasion (large body size) is not selected under favourable conditions.

7.7 Life history trade-offs and stress resistance

Life-history theory does not directly incorporate variation in resistance to environmental stress and is generally only concerned with stress responses involving evasion. We will consider the question of how variation in stress resistance may influence genetic trade-offs between life-history traits.

It seems likely that any trade-off between life history traits will tend to disappear under stressful conditions because performance on most life-history traits will primarily be determined by the stress resistance of individuals. Plants that are able to grow under conditions of drought or nutrient stress are also likely to have higher reproductive output and survival under these conditions. Animals that can survive starvation conditions because of a low metabolic demand are also likely to have higher rates of reproduction when their food supply is intermittent. If there is little genetic variation for stress resistance in a population, correlations between life-history traits may not be evident under stressful conditions, because mechanisms for coping with an environmental stress override individual differences in allocation to life-history traits. Genetic variation for stress resistance could lead to positive genetic correlations between life-history traits under stressful conditions because animals and plants that survive and/or continue growth when they are under stress are also likely to have higher overall reproductive outputs. Genes that increase stress resistance may therefore be associated with increased performance on most life-history traits when tested under stressful conditions.

Correlations between life-history traits may also be influenced by trade-offs between stress resistance and fitness under favourable conditions, particularly if the costs associated with resistance are large. If stress resistance genes have deleterious effects on a range of fitness traits then genetic variation for stress resistance may obscure trade-offs among life-history traits. Correlations between life-history traits could therefore be influenced by levels of genetic variation for stress resistance which will, in turn, depend on the history of selection in a population.

These considerations suggest a wider framework for examining genetic correlations between life-history traits. We need to know what effects stressful conditions have on processes that are responsible for trade-offs between life-history traits, such as those that mediate the allocation of resources to growth, reproduction and longevity. A starting point that may be relevant to many stresses is to view resistance as acting prior to resource allocation, such as via 'environmental filtering' (Section 3.3). Many stress resistance traits act as filtering mechanisms, helping to maintain metabolic homoeostasis by minimizing the impact of an environmental stress at the metabolic level. For example, wing patterns, and other pattern and colour polymorphisms help to regulate internal temperature; chemical stressors such as high concentrations of salt and other toxins are filtered by resistance mechanisms that keep them away from the cytoplasm by sequestration or detoxification. Even a reduction in metabolic rate may be considered as a filter because it minimizes the impact of an energetic shortage on metabolic processes, although 'filtering' in this case is probably exerted at the level of intermediary metabolism rather than the environmental interface. Organisms under stress are more likely to be subjected to large changes in substrate concentrations and to other forms of metabolic disturbance that will reduce the flow of resources available for allocation to life-history traits. The effects of stress on life-history correlations will depend on the consequences of this reduction. For example, individual differences in the overall availability of resources because of variation in environmental filtering may have a larger influence on the expression of life-history traits than patterns of resource allocation, in which case variation in life-history traits will reflect variation in stress resistance rather than life-history trade-offs.

Do correlations between life-history traits depend on environmental conditions and the history of selection in a population? A spate of papers on genetic correlations among life-history traits has recently appeared, often with conflicting results. Unfortunately most of these studies have not considered correlations between traits over a range of environments, and have usually considered only one population. There is some evidence that correlations are not constant and vary between populations (e.g. Lofsvold 1986; Dingle *et al.* 1988; Cohan and Hoffmann 1989) but these differences have generally not been related to variation in the selection history of populations.

Giesel and co-workers (Giesel and Zettler 1980; Giesel *et al.* 1982; Murphy *et al.* 1983; Giesel 1986) found positive genetic correlations between development rate, reproduction, and longevity using inbred or partially inbred lines of *D. melanogaster*. Correlations were usually positive under stressful temperatures or stressful nutritional conditions as well as under more optimal conditions, although correlations showed some changes with temperature. This work has been criticized because of the possibility that positive correlations were the result of inbreeding. Deleterious alleles could be fixed by inbreeding, causing some of the lines to perform poorly on all traits (Rose 1984). In addition, some of Giesel's experimental designs tested differences among full sibs, allowing for maternal and non-additive genetic effects which could mask the genetic correlations (Reznick 1985). Finally, stocks used in these experiments were tested soon after they were taken from nature, and the laboratory environment probably represents a stress for most organisms when they are first encountered (Kohane and Parsons 1986; Clark 1987). All of the genetic correlations obtained in these experiments may therefore have been scored under stressful conditions. Positive correlations between life-history traits have also been found in other studies where stocks from the wild were tested soon after being brought into the laboratory, such as Bell's (1984*a,b*) experiments with freshwater invertebrates. These findings suggest that stressful conditions may generate positive genetic correlations among life-history traits.

In contrast to the experimental results of Giesel and others, Rose and Charlesworth (1981) and Luckinbill *et al.* (1984) found evidence for negative genetic correlations between early fecundity and longevity in *D. melanogaster*. The former used a laboratory-adapted population, while the latter obtained correlations from a selection experiment, when there was an opportunity for stocks to adapt to laboratory conditions. Service and Rose (1985) examined the correlation between starvation time (related to longevity, see above) and early fecundity in a 'novel' environment consisting of a different type of *Drosophila* medium, rearing temperature, and light regimen. The negative correlation between these traits (-0.91) in the 'Standard' environment on which the flies were usually raised was reduced to -0.45 in the novel environment. They suggested that the correlation is more positive because genotypes that are fortuitously adapted to the novel environment would be expected to have generally enhanced fitness. An alternative explanation consistent with the arguments presented here is that the more positive correlation is associated with a stressful environment rather than a novel environment. Such stressful conditions may not be novel but may be commonly encountered by flies in natural situations.

Scheiner *et al.* (1989) considered the genetic correlation between longevity and fecundity in a mass-bred population of *D. melanogaster* at 25°C and at 19°C. Negative correlations between life-history traits were present at 25°C but not 19°C, indicating that the correlations were temperature-dependent. As 25°C is considered optimal for this species (Tantawy and Mallah 1961; Parsons 1977), trade-offs were apparent under optimal conditions. However, 19°C does not represent a stressful temperature for *D. melanogaster* and it would be interesting to see whether the correlation becomes positive under more extreme temperature stress, since there is evidence to suggest that genes affecting temperature resistance may only be expressed under extreme conditions: Tantawy and Mallah (1961) showed that *D. melanogaster* from a warm environment were larger than those from a cooler region only at temperatures of 28°C or higher.

These experiments indicate that genetic correlations can change with the environment, but they are inadequate to test whether environmental stress generates positive correlations among life history traits. Nevertheless, this prediction may help to explain the common finding that genetic correlations among life history traits tend to be positive when stocks are tested in the laboratory soon after they are brought in from the natural environment.

7.8 Summary

Costs associated with increased stress resistance may arise because of patterns of resource allocation, constraints associated with the presence of resistance mechanisms and a reduced ability to utilize resources under optimal conditions because of resource conservation.

Increased resistance to specific chemical stresses such as herbicides, pesticides, and heavy metals is often associated with deleterious fitness effects under optimal conditions.

Agricultural and laboratory studies indicate that genotype–environment interactions are widespread, and there are a few cases where negative genetic correlations occur between performance in stressful and optimal environments. Indirect evidence from natural populations supports the possibility of genetic trade-offs between environments, but more data collected under demonstrably extreme environments are needed to determine consequences at the evolutionary level.

Variation in metabolic rate and growth rate provides one approach for investigating the physiological basis of trade-offs between stress resistance and fitness under favourable conditions. Such trade-offs may not be apparent when fitness traits are correlated with increased metabolic efficiency which may be associated with heterozygosity levels at enzyme loci.

Many aspects of life-history theory are concerned with stress evasion rather than resistance. Mechanisms for stress evasion may be associated with costs because of resource allocation or constraints imposed by traits required for evasion, or because of life-history trade-offs in heterogeneous environments.

Genetic correlations between life-history traits may tend to become positive as stress is increased because variation in these traits will ultimately reflect variation in stress resistance. Large deleterious effects associated with some resistance mechanisms may tend to make life-history correlations more positive under favourable conditions.

8. Stress, species margins, and conservation

In the preceding chapters we have demonstrated and analysed the features of genetic variation for stress responses within populations. Here we link genetic variation with the ecological considerations of the first two chapters. A number of theories of range expansion will be reviewed, emphasizing the role of genetic variation for stress response traits in limiting species distributions. The preservation of genetic variation for these traits is also considered. Adaptation to environmental stresses is particularly important in the long term conservation of populations and species. An understanding of genetic changes underlying stress responses may clarify the ability of organisms to survive rapid environmental changes resulting from human activities, including the global warming predicted under the greenhouse scenario, the increasing impact of chemical pollutants, and the increased perturbations of the physical environment following habitat destruction.

8.1 Range expansion and marginal populations

Organisms are often not well suited to habitats where physical factors are extreme. Plants from stressed environments often perform relatively better under more favourable conditions: bog plants tend to grow better when their soils are drained (Crawford 1983); those normally restricted to areas of low fertility show increased growth when transferred to areas of higher fertility (e.g. Bradshaw *et al.* 1965). The salt marsh plant, *Halimione portulacoides*,

> . . . seems totally unfit for the habitat in which it grows: it prefers aerated soils . . . although large parts of the salt marsh soils are anaerobic; it is sensitive to inundation . . . whereas saltmarshes are flooded regularly; it grows best at low salt concentrations . . . although salinity of the salt marsh soil is usually much higher; its frost hardiness is low . . . and it occurs in salt marshes of the temperate climate (Huiskes and Stienstra 1985).

Indeed many genecological studies indicate that ecotypes from harsh environments grow more luxuriantly under garden conditions than in the field (Bradshaw 1965); plants are therefore limited in their ability to respond to the extremes of physical factors normally encountered in their habitats.

The inability of organisms to respond to harsh environmental conditions is also suggested by the importance of extremes of particular climatic variables in determining species ranges (Section 2.3). Range expansion may be limited by the inability of many populations at margins to increase their stress resistance or their inability to evade stressful conditions. These margins do not necessarily represent geographic barriers or abrupt environmental changes (Mayr 1963); they may be determined by a complex of environmental factors, as discussed in Section 2.3. The poor performance of plants in harsh conditions and limits of species distributions contrast with the abundant evidence for genetic variation in stress response traits within populations discussed in Chapters 4 and 5. Why then are species not more able to adapt to harsh conditions and why do they not extend their ranges? These questions suggest that explanations for limits to adaptation should be sought initially at the genetic level and then integrated with ecological observations on the factors limiting range expansion.

Most investigations of species borders have focused on interspecific comparisons: species borders are often explained in terms of one species being relatively less resistant to a particular environmental stress than another. For example, Carmi-Winkler et al. (1987) investigated the distribution of chukars (Alectoris chukar), which are widespread but do not occur in areas of extreme aridity. Chukars spend much of the hot part of the day resting in the shade, restricting their foraging time to avoid heat stress. Sand partridges, which are more resistant to heat than chukars, maintain their body temperature at an ambient temperature of at least 51°C, compared to 43°C for chukars (Frumkin et al. 1986); predictably, they occur under more arid conditions than chukars. The association between stress resistance and the distribution of Drosophila species outlined in Fig. 2.5 represents another example of this approach.

While such studies provide information about physiological variables that might underly limits to species distributions, they do not directly address the question of why a species border occurs in a particular place and why it is stable. The relatively higher heat resistance of sand partridges does not explain why chukars have lower heat resistance. Why have chukars not evolved higher heat resistance and expanded their range? Species are not static entities, since the genetic constitution of populations changes continuously in response to selection, genetic drift, mutation and gene flow. An understanding of why a border occurs where it does will require an understanding of how these processes limit further evolutionary change.

Hypotheses for species borders which emphasize the genetic level have been considered, particularly by Mayr (1963), Antonovics (1976) and in an unpublished thesis by Ward (1985); these are listed in Table 8.1, with additional possibilities. We discuss these hypotheses with the assumption that inability to respond to environmental stresses is of primary importance in limiting species ranges. Factors such as predation, parasitism and competitive ability may also be important in determining species borders. These are not directly addressed, except to note that biotic factors will often interact with the environmental factors.

Table 8.1. Hypotheses about limits to the range expansion of marginal populations when ranges are determined by environmental stresses (includes hypotheses from Mayr 1963, Antonovics 1975, and Ward 1985).

1. Low levels of genetic variation in stress response traits as a consequence of directional selection and physiological limits.
2. Low overall levels of genetic variation because of population structure.
3. Low levels of genetic variation in stress response traits expressed under stressful conditions but not under optimal conditions.
4. Changes in several independent characters required for range expansion.
5. Negative genetic correlations for performance under stressful and favourable conditions.
6. Negative genetic correlations among fitness traits expressed under stressful conditions.
7. Swamping of genotypes favoured in marginal environments because of gene flow from central populations.
8. Inbreeding as a result of small marginal population size.
9. Heterozygotes favoured under extreme conditions.

8.1.1 Hypotheses based on reduced genetic variation

If there is little genetic variation in stress response traits in marginal populations, the range of species cannot expand because populations cannot respond to selection pressures at the species border. The first three hypotheses in Table 8.1 are concerned with reduced levels of genetic variation in marginal populations. In hypothesis 1, populations may have been under directional selection for stress responses in the past and reached a limit because there is little genetic variation left segregating in the base population. This limit may reflect physiological or biochemical constraints on further evolutionary change. In hypothesis 2, the geographical isolation of marginal populations may have resulted in substantial genetic drift, reducing the overall level of genetic variation in a population, including that for stress response traits. Species will often occur at low abundance in marginal populations and because these populations are poorly adapted to their environment they are likely to fluctuate in response to climatic changes. These factors will decrease the effective population size of marginal populations and therefore increase the impact of genetic drift. Rare alleles that may be important in stress responses are less likely to arise by mutation in small populations and may be lost in populations that show large fluctuations in size (Nei *et al.* 1975). In hypothesis 3, genetic variation for stress responses may be present in marginal populations but not expressed in an adaptive way in the adverse conditions to which these populations are exposed, leading to a reduction in heritability of adaptedness under stressful conditions (Ward 1985). There are several ways in which this could occur. Ward (1985) suggests that the adaptive value of an organism's genotype is only expressed under the conditions in which it has been selected. Species margins are often associated with environmental extremes that are uncommon (Section 2.3), so selection for traits of adaptive value under these conditions may occur rarely. Genetic differences in stress response traits may therefore not be expressed under marginal conditions. Blum (1988) argued that marginal conditions will reduce the heritability of fitness-related traits in plants because small fluctuations in environmental effects will lead to large changes at the phenotypic level, with the result that most of the phenotypic variance is environmental. Another possibility is that exposure to extreme conditions triggers resistance mechanisms that are common to all individuals, and this may override the expression of genetic variation for traits influencing fitness in the absence of a stress.

These hypotheses make somewhat different predictions. Hypothesis 2 predicts an overall decrease in genetic variation in marginal populations, regardless of whether or not the type of variation being considered is relevant to stress responses in marginal environments. Hypothesis 1 predicts that genetic variation for stress response traits will be less in marginal populations than in populations at the center of species distributions. This prediction holds regardless of the environment in which the genetic variation is characterized. In contrast, the third hypothesis predicts that genetic variation for stress response traits will be reduced only under conditions experienced by marginal populations, regardless of whether individuals are obtained from marginal populations or from central populations. Levels of genetic variation in the marginal and central populations may be similar under favourable conditions.

Such hypotheses about species borders are difficult to evaluate because of a paucity of empirical data. Allozyme data provide some evidence for decreasing levels of overall genetic variation in marginal populations compared to central populations in some species (e.g.

Vrijenhoek *et al.* 1985; Jaenike 1989) supporting hypothesis 2, although this relationship does not hold for other species (Myers and Sabath 1980). It is not clear how low levels of allozyme variation can be translated into heritabilities for stress response traits, particularly as only some allozyme polymorphisms have been related to stress resistance (Section 3.3). Genetic variation, measured by chromosome inversions, has also been related to the marginality of populations, particularly in *Drosophila*. In 15 of 16 *Drosophila* species, marginal populations (usually defined by geographic criteria) show a paucity of chromosome inversions compared to central populations (reviewed in Soulé 1973; Brussard 1984). These differences are not matched by allozyme variation in the same species (Soulé 1973), suggesting that inversion levels do not represent general estimates of the overall genetic variation in a population. Other explanations may account for the low level of inversion polymorphism in marginal populations, as discussed in Brussard (1984).

Genetic studies of stress response traits discussed in earlier chapters are not particularly useful in evaluating these species border hypotheses, because the studies were usually carried out with laboratory populations originating from favourable environments rather than populations from species margins. Genecological studies of plant populations are also of limited relevance, since these have focused on comparisons between populations occupying different environments that are not at species margins. Heritability estimates for stress response traits have generally been obtained under one set of laboratory conditions, and we know little about the heritability of these traits in natural populations from favourable and unfavourable conditions.

One of the few cases where the importance of heritable variation in local range expansion has been considered is that of the heavy metal resistance of plant species growing on mine tailings, where genetic variation may be limiting. Heritable variation for resistance was found in individuals from non-mine populations of species that occurred on mine tailings, but resistant individuals were generally absent in species that did not occur on tailings (Ingram, cited in Bradshaw 1984). The evolution of heavy metal resistance may therefore have depended on the presence of genetic variation in the base populations, rather than on rapid genomic changes induced by stressful conditions (Section 5.1). There is also evidence that the lack of genetic variation may limit further adaptation to heavy metals. In the moss, *Funaria hygrometrica*, heritabilities for resistance to copper were high in three populations where heavy metals were absent, but low in a population originating from a copper mine (Shaw 1988). Both of these findings support hypothesis 1 rather than hypothesis 3; differences between non-mine populations and the marginal mine populations were expressed under the same environmental conditions. Hypothesis 2 is probably not important in this case because resistant populations occur adjacent to populations in uncontaminated sites, so that there is no geographic isolation that could result in a low population size.

The absence of genetic variation may have limited the evolution of resistance to other chemical stresses. For example, resistance to phenoxy herbicides in weeds has not evolved in over 40 years, despite the demonstration of resistance in crop plants and the ability of weeds to evolve resistance to other herbicides (Gressel 1984). Similarly, many plant species are unable to recover from smelter emission damage, while other species have successfully extended their range after an initial population decline (Hutchinson 1984). Both cases suggest insufficient genetic variation in the base population of species to counter the effects of the chemical stress (hypothesis 1).

There are few genetic comparisons between the responses of central and marginal populations to climatic stresses of importance in determining species distributions. One example where such genetic variation may have been reduced in marginal populations because of directional selection (hypothesis 1) is desiccation resistance in *D. melanogaster* populations from coastal Australian sites (Stanley and Parsons 1981). Populations from Melbourne and Darwin have higher desiccation resistance than populations from Brisbane and Townsville, as predicted by climatic considerations (Section 4.2). Populations from the more stressful environments also have reduced levels of genetic variation, as indicated by the variance estimates among isofemale strains (Table 5.7). Climatic selection may have decreased the amount of genetic variation in the Melbourne and Darwin populations. Similar trends were found for ethanol resistance which is genetically correlated with desiccation resistance (Section 6.6). However, there is ample genetic variation for desiccation resistance in the Melbourne population (Hoffmann and Parsons 1989*a*), which suggests that a low level of genetic variation for this trait is not limiting range expansion by *D. melanogaster*. These types of studies are needed in species less widespread than *D. melanogaster* so that climatically marginal populations can be more readily identified.

Evidence for decreased heritability under marginal conditions was obtained by Agnew (1968) in populations of the herb, *Lysimachia volkensii*. Plants originating from populations along three transects extending from a central area of the range to marginal areas were grown in the same environment and characterized for variation in floral measurements and fruit diameter. The variance for these characters between individuals was highest in central populations and decreased markedly towards the periphery of the species range along the three transects. Even so, the relevance of the measured traits to environmental stress responses at the species margins is not known. In contrast, Scheiner and Goodnight (1984) did not find an association between the heritability of morphological traits and marginality in five populations of the grass, *Danthonia spicata*. The authors suggested that this may reflect the short time (seven generations) which separated the most marginal population they examined from the most central population.

Relevant experimental data have also been collected in experiments testing one of the explanations for the low level of inversion polymorphisms in marginal *Drosophila* populations. Carson (1959) proposed that inversions provide a form of heterozygote buffering which improves performance in a variety of niches available in the central population, while marginal populations are too small and inbred to be buffered in this manner. In addition, he proposed that the low frequency of inversions in marginal populations increases recombination rate, allowing rapid genetic change in response to environmental fluctuations. This hypothesis predicts that the selection response should be greater in marginal populations than in central populations that are polymorphic for inversions, in contradiction to the above hypotheses. Carson (1958) artificially selected for motility for six generations in isofemale lines of *Drosophila robusta* originating from a marginal and a central location. Eight of the 10 lines from the marginal location responded to selection, whereas seven of the 15 lines from the central location showed a response, suggesting the possibility of more genetic variance for motility in the marginal population, although these proportions do not differ significantly by a contingency test. Tabachnick and Powell (1977) also tested this hypothesis using two island populations of *D. willistoni* that were relatively monomorphic for inversions and two mainland populations that were polymorphic for inversions and came from near the center of *D. willistoni*'s range. They selected lines by culturing flies on medium with increasing levels of different chemical stresses and compared the number of cultures of each population surviving repeated stress periods. More cultures from the monomorphic populations survived all chemical stresses than from the polymorphic populations (Table 8.2), in agreement with Carson's findings.

These experiments suggest that marginal populations do not have lower levels of genetic variance than central populations. However Carson's central and marginal populations were not replicated, so his results could reflect location differences unrelated to the central–marginal dichotomy. The selection responses of *D. melanogaster* for increased knockdown resistance (Cohan and Hoffmann 1986) illustrate that populations do not necessarily respond in the same way to the same selection pressure. It is noteworthy that the two central and marginal

Table 8.2. Survival of monomorphic and polymorphic populations of *Drosophila willistoni* exposed to four chemical stresses. (After Tabachnick and Powell 1977.)

Chemical	No. of initial cultures per population	Number of surviving cultures			
		Monomorphic		Polymorphic	
		Pop 1	Pop 2	Pop 1	Pop 2
Propionic acid	24	22	7	1	0
NaCl	22	22	18	3	14
KI	25	25	19	6	11
CaSO$_4$	25	14	5	4	0

populations responded differently for some traits in Tabachnick and Powell's experiment. Another limitation of these experiments is that the relevance of the selected traits to adaptation at the ecological margin is unknown. Tabachnick and Powell suggested that resistance to chemical stresses may be ecologically relevant to selection in marginal populations, but the discussions in Chapters 4 and 5 suggest that responses to chemical stresses are often specific, and may not be related to traits such as resistance to climatic stresses that commonly correlate with species borders. These results therefore do not provide a test of hypothesis 1, but provide evidence against hypothesis 2 in agreement with the allozyme data for *Drosophila* species (Soulé 1973).

Hypothesis 3 requires that genetic variance in stress response traits should be reduced as stress levels increase. In animals the genetic variance often increases rather than decreases as levels of stress become more severe (Section 5.4). However, many of the studies have involved animals cultured under optimal conditions, making it difficult to relate these findings to the hypothesis that heritable variation is reduced when animals live in marginal conditions. In contrast, comparable plant studies have been carried out with individuals cultured under stressful conditions; some of these studies found that heritabilities for yield decreased with increasing stress levels (Blum 1988), as predicted by several of the hypotheses discussed in Section 5.5. In addition, differences between susceptible and resistant genotypes may be reduced under stress when the environment induces plastic changes that increase stress resistance more in susceptible genotypes. Some evidence for this is provided by the association between phenotypic plasticity and stress resistance discussed in Section 4.6, which suggests that genotypes with lower inherent resistance levels are often more plastic, so that their morphology and physiology under extreme conditions approaches that of more resistant genotypes. Further evaluation of hypothesis 3 will require genetic experiments involving a series of populations extending from the species border, and involve the characterization of genetic parameters for traits important in determining species distributions under a range of favourable and unfavourable environmental conditions. These experiments would parallel the comparisons of crop plant genotypes under a range of environmental conditions that are commonly used to evaluate yield potential under diverse environmental conditions (Section 7.3).

8.1.2 *Other hypotheses*

Several hypotheses shown in Table 8.1 do not involve low levels of genetic variation in marginal populations. Hypothesis 4 proposes that range extension is restricted by the requirement for simultaneous changes in several characters which may be needed to respond to multiple environmental changes at species borders, or to an environmental stress requiring a complicated resistance mechanism. Complexes of genes for range extension were emphasized by Wallace (1959) and Mayr (1963). Individuals with the genotype fitted for a new environment may occur rarely, even when there is genetic variation, particularly if genes increasing stress resistance are present at low frequencies in the population. The importance of this hypothesis will depend on the types of environmental stresses at species boundaries: it may be particularly relevant when changes at borders are large and abrupt, and involve several environmental factors. Several climatic variables are often associated with species borders (Section 2.3) suggesting that range expansion is not limited by a single factor, except in some cases of local variation such as plant populations limited by heavy metals or soil moisture. Nevertheless, some individuals in a population may show increased resistance to a range of stresses because genetic variation for resistance to different stresses may be correlated (Section 6.6). Species might be expected to adapt more readily to some combinations of stresses than other combinations. For example, adaptation to desiccation and starvation may be more likely than adaptation to temperature fluctuations because cold and heat resistance are probably not correlated except perhaps via heat-shock proteins.

A fifth hypothesis is that range extension does not occur because of genotype–environment interactions. If there is a negative genetic correlation between performance in two environments, then increased performance in one environment means decreased performance in the other: this restricts evolutionary change in populations that encounter both environments. The possibility of trade-offs between stress resistance and performance under more favourable conditions (Chapter 7) suggests that genes increasing resistance may decrease in frequency when conditions become favourable. Boundaries where climatic conditions fluctuate and extremes occur rarely may, therefore, not extend readily, because resistant genotypes are not continuously favoured.

Trade-offs could involve different traits rather than environments (hypothesis 6) as argued by Antonovics (1976). For example, competitive ability could be favoured in marginal populations where interspecific competition is intense and this trait may show a trade-off with stress resistance (Section 7.4). Grant (cited in Antonovics 1976) found some evidence for negative phenotypic correlations between different fitness-related traits in a population of *Anthoxanthum odoratum* from a marginal environment that might reflect this type of limit to selection, although genetic correlations were not determined. The commonly observed decrease in fitness that accompanies artificial selection for quantitative traits in *Drosophila* and other organisms suggests that responses to directional selection will often be limited by genetic interactions with fitness-related traits.

A seventh hypothesis, emphasized by Mayr (1963), is that range extension is limited by gene flow from a large, non-marginal population where stress responses are not selected to the same extent. Marginal populations tend to be small because they occupy harsh environments; immigrants may therefore contribute a large proportion of the gene pool each generation, limiting the rate of increase of locally selected genes. In the extreme case, marginal populations sporadically become extinct and the area may be recolonized by immigrants from central populations. This appears to occur in pied flycatchers in a marginal area of Finland where the population cannot maintain itself in poor breeding years (Jarvinen and Vaisanen 1984). Gene flow could be particularly effective in limiting adaptation to marginal conditions if there is a trade-off between stress resistance and performance under optimal conditions (hypotheses 7 and 5/6 combined). Resistance genes may gradually increase in a central population if there is migration from the marginal population, but this is less likely when there is a trade-off and resistance genes are selected against in the central population. Gene flow may also be particularly effective in retarding the evolution of genotypes adapted to multiple environmental factors (hypotheses 7 and 4 combined).

Inbreeding arising from the small size of marginal populations could mean that deleterious recessive alleles decrease the fitness of individuals (hypothesis 8). This should be reflected in comparisons of fitness traits between central and marginal populations, but the few relevant *Drosophila* studies do not provide supporting evidence. Pfreim and Sperlich (1982) compared the viability of extracted chromosomes in populations of *Drosophila subobscura* and found that the mean viabilities of non-lethal homozygous and heterozygous chromosomes were higher in northern marginal populations in Europe than in central populations. Viabilities should have been lower in the marginal populations if inbreeding was causing the fixation of deleterious recessive genes. This hypothesis may be more relevant to species that are less vagile than *Drosophila*, when inbreeding will be more intense.

A ninth hypothesis is that heterozygotes are the fittest genotypes in extreme environments (Parsons 1971). This follows from the observation that heterozygous genotypes from crosses between different populations or strains tend to have an advantage under stressful conditions (Section 5.3). If marginal environments always favour heterozygotes, then genetic variation for stress resistance will persist in marginal populations. This hypothesis predicts that marginal populations will not respond greatly to directional selection for increased stress resistance, and that crosses between lines from marginal populations differing in resistance will produce overdominance. Some evidence supporting this has been found in *Drosophila subobscura* populations whose increased viability caused by heterozygosity is greater in marginal populations than in central populations (Sperlich *et al.* 1977; Pfriem and Sperlich 1982), although it is not clear how viability on laboratory medium translates into responses to stresses encountered in the more variable environments of the wild. Evidence against this hypothesis includes the high degree of additive genetic variance for stress resistance within populations (Section 5.4), and the low level of inversion polymorphism in marginal *Drosophila* populations. If heterozygous genotypes had an advantage then inversion heterozygotes would be favoured in marginal populations and polymorphism would be maintained. Instead, it has been argued that inversion heterozygotes may be favoured in central populations rather than marginal populations, because of the diversity of ecological niches or because chromosome heterozygotes have an advantage under the high density conditions experienced in central populations as a consequence of their superior competitive ability (see Brussard 1984). These ideas have not been directly tested. In general, the role of heterozygote advantage in maintaining inversion polymorphisms is not clear (Anderson *et al.* 1986). Processes such as density- and frequency-dependent selection have recently been shown to operate on inversion polymorphisms in *D. pseudoobscura*, at least in the laboratory (Anderson *et al.* 1986), and these will be more important in central populations where density is higher. However there is also evidence that inversion heterozygotes perform better under stressful conditions than homozygotes (Section 5.3) and show increased resistance to desiccation stress (Thomson 1971).

The above hypotheses have been discussed in terms of stress resistance traits, but it should be emphasized that they will also apply to stress evasion traits. Species borders may be influenced by stress evasion via dispersal ability rather than adaptation to conditions at species margins. In a study of genetically correlated wing polymorphisms in arthropods, Southwood (1962) argued for higher migratory ability in marginal unstable habitats at the edge of the range of species, which could be a reflection of activity differences and variation in the level of migrational selection according to habitat (Parsons 1963). Poor dispersers are expected to disappear from small and isolated habitat fragments that may be characteristic of species margins (Den Boer 1985), and increased dispersal ability should be selected at species margins. There is evidence that dispersal ability can influence species distributions: Turin and Den Boer (1988) showed that carabid beetle species which are poor dispersers have contracted in range in the Netherlands, whereas the distribution of beetles which are good dispersers has remained relatively stable. Dispersal ability may also be important in the local distribution of tree species, as indicated by palaeoecological studies of range limits (e.g. Woods and Davis 1989).

8.1.3 Indirect evaluations of species border hypotheses

The importance of these hypotheses cannot be adequately evaluated in the absence of much genetic data from marginal and central populations. However some indication of the importance of genetic variation in limiting range expansion can be obtained indirectly from non-genetic data. Transplant experiments can indicate whether or not species from marginal environments survive beyond the species boundary; these results can be interpreted as evidence for the presence or absence of genetic variation in a marginal population. For example, Jefferies *et al.* (1983) considered a marginal population of the annual plant, *Salicornia europaea*, which was confined to the warmer south-facing sites on the shores of Hudson Bay. Seeds transplanted to a north-facing site showed poor or no germination, and the seed production of mature individuals was low; genetic variability in the marginal population was probably insufficient to permit survival of any individuals in the colder habitat.

Another indirect method of demonstrating the limits of genetic variation is to show that the borders of different species can be predicted by the same underlying physiological factor: a physiological constraint common to the species may, in turn, reflect a genetic constraint. The northern distribution of bird species in North America, which can often be related to isotherms for the mean minimum January temperature (Root 1988*a*), is an example of this. Root (1988*b*) found that the resting metabolic rate of 14 bird species at this minimum temperature was about 2.5 times the basal metabolic rate of these species. This relationship also held for 36 additional species whose metabolic rates were determined indirectly from the allometric relationship between metabolic rate and body size. The raised resting metabolic rate arises because energy is expended to counter the effects of low temperatures. These results suggest that species ranges are determined by the amount of energy needed to keep warm, regardless of variation in body size or habitat. Genotypes that enable individuals to function beyond this limit may not exist because of physiological constraints. Such findings are consistent with hypothesis 1, and perhaps with hypothesis 4. It seems unlikely that the effects of population structure (hypothesis 2) and gene flow (hypothesis 7) on limiting range expansion would be the same in unrelated species. It also seems unlikely that the different marginal environments would have the same influence on the genetic variance (hypothesis 3).

8.1.4 Concluding remarks

In general, there is insufficient information to decide between these alternative hypotheses, although some genetic data and indirect evidence suggest that genetic variation at species margins may be limiting, and/or fitnesses of genotypes may become restrictive at margins. The relative importance of these hypotheses will depend on factors such as the magnitude of the fitness trade-offs for stress resistance genes, and the genetic correlations between different stress response traits. More empirical work on marginal populations is required, particularly in examining genetic variation for traits related to climatic stresses and the metabolic costs associated with these traits. Possible candidates for such studies are rainforest species of plants and insects, which have limited distributions and are likely to be particularly susceptible to climatic extremes. Relevant information will also come from genetic studies of species that have undergone a recent range expansion. Hypotheses involving a lack of genetic variation would be implicated if genotypes from populations in newly colonized habitats are not found in the original parental populations, while hypotheses involving gene flow would be implicated if species rapidly expanded their range after the size of the central population in a favourable environment is reduced. This could occur when species colonize a new area with favourable conditions: wild rabbit populations in Australia show a range of adaptive physiological and morphological differences between climatic regions that have evolved since the introduction of 24 rabbits from Europe (Williams and Moore 1989). Genetic comparisons between these populations and European populations could indicate the relative importance of this bottleneck as opposed to genetic constraints in limiting range expansion. The question of range expansion is relevant to genetic conservation under global temperature shifts (see below).

8.2 Stress and conservation

In *The genetical theory of natural selection,* Fisher (1930) wrote:

If therefore an organism be really in any high degree adapted to the place it fills in the environment, this adaptation will be constantly menaced by any undirected agencies liable to cause changes to either party in the adaptation.

The genetics of environmental stress resistance, particularly with reference to stresses caused by man, was little considered at that time. However, Parsons' (1974) review of the genetics of stress resistance in *Drosophila* included references to various chemicals in the environment derived from the activities of man, and environmental stresses are now recognized as being complex and widespread. The atmosphere contains increasing concentrations of CO_2 and less abundant atmospheric gases including chlorofluorocarbons, and substantial exchanges of material between terrestrial systems and the atmosphere occur (Mooney *et al.* 1987) which are likely to cause major depletions in stratospheric O_3 (Cicerone 1987). While it is difficult to forecast future trends, these alterations have collectively been predicted to increase world temperature by up to 6°C within the next century—perhaps more for localities distant from the equator (Schneider 1989). Compared with prehistoric changes of a similar magnitude (Peters and Darling 1985), this 'greenhouse effect' is an exceedingly rapid change.

These major climatic changes and increases in pollution are likely to impose environmental stress on much of the world's biota in the near future. The successful conservation of species in natural habitats and in zoos (with subsequent release into natural habitats) clearly depends on the ability of organisms to adapt to environmental change. Genetic variation for stress response traits is a prerequisite in situations where organisms cannot escape stressful conditions by moving into new habitats. This will become increasingly critical as habitats become more restricted and fragmented due to human activities, making it more difficult for organisms to move to favourable habitats. The conservation of genetic variation for stress response traits is therefore important in ensuring long term survival of species, and we now consider factors that are likely to affect this process.

8.2.1 Usefulness of current approaches

Discussions of conservation genetics usually consider the significance of population size, bottlenecks, drift, and inbreeding depression on levels of genetic variability within a population or species (e.g. Frankel and Soulé 1981; Schonewald-Cox *et al.* 1983; Soulé 1987; Simberloff 1988). The development of island biogeographic theory by ecologists has also led conservation biologists to examine refuge design for particular ecosystems and regions. More recently, demographic models for the persistence of populations buffeted by random catastrophic events have been developed, and the effects of population extinction on genetic variation have been considered (Soulé 1987; Simberloff 1988).

Conservation genetics has focused primarily on maintaining the overall level of genetic variation in populations, as measured by protein electrophoresis; genetic variation for quantitative morphological traits has been considered to a lesser extent. Minimal attempts have been made to identify and study genetic variation in the types of traits that may be important in the future survival of a species under changing environmental conditions (Parsons 1989*b*). Conserving the maximum amount of allozyme variation has been justified on the basis of the relationship between multilocus heterozygosity and fitness (e.g. Frankel and Soulé 1981), which was discussed in Section 7.5. If fitness is always positively associated with heterozygosity levels, then maximizing levels of heterozygosity in natural populations may help the future survival of a species.

However the association between multilocus allozyme heterozygosity and fitness may be weak or non-existent in some species, casting doubt on the usefulness of allozyme heterozygosity as a general indicator of overall fitness. The explanation for multilocus heterozygote advantage presented in Section 5.5 does not require inherent heterozygote advantage, suggesting that some individuals with low heterozygosity levels have superior performance over highly heterozygous individuals. In any case, many widespread and abundant species have low levels of allozyme heterozygosity (Nevo 1978), emphasizing that this is not necessarily related to population persistence. Data from introduced species indicate that allozyme heterozygosity is not a useful predictor of whether or not a species will be a successful colonizer (Myers and Sabath 1980). No association between multilocus allozyme heterozygosity and performance has been generally established for stress response traits, suggesting that the conservation of high allozyme heterozygosity levels does not guarantee the conservation of the genetic variation which is important in adapting to stressful periods. Lande (1988) has emphasized that different kinds of genetic variation are not necessarily proportional, and schemes to maximize the conservation of one type of variation can have little impact on variation of another type.

In conservation genetics much emphasis has been placed on the

minimum viable population size needed for the persistence of a population. Estimates of this effective population size (N_e) stem from two
theoretical requirements: the minimum size needed to avoid deleterious
effects associated with inbreeding, and the minimum size needed to
overcome the loss of genetic variation by drift. Franklin (1980) suggested that the former should be about 50 and the latter around 500.
Empirical data indicate that inbreeding depression is not a problem
when populations are maintained at an effective size of about 50. The
figure of 500 represents an equilibrium when genetic variance of a
quantitative trait lost by drift is balanced by the genetic variance
generated by mutation with the maintenance of a heritability of around
0.5. Franklin's analysis assumes:

(1) that the genetic variance is additive;

(2) that these evolutionary processes are the only factors affecting
    genetic variation;

(3) that the environmental variance is similar to the genetic variance
    (Lande and Barrowclough 1987);

(4) that mutation for quantitative traits is 10^{-3}; this estimate is based on
    values obtained in *Drosophila* experiments on bristle number, a
    quantitative trait known to be under stabilizing selection (Kearsey
    and Barnes 1970).

Lande and Barrowclough (1987) considered the case of a trait under
stabilizing selection and suggested that the same mutation rate can
maintain intermediate heritabilities when the effective population size is
500 or greater.

How useful are these estimates for stress response traits? The
inbreeding estimate is based on the experience of animal breeders which
indicates that domestic animals can tolerate an inbreeding level of 1 per
cent per generation without much influence on traits such as yield and
productivity. No experiments have been directed at determining depression of stress resistance by inbreeding, but inbreeding effects may be
more severe under stress; heterosis from crosses between inbred lines
tends to increase as environmental conditions become harsher (Section
5.3). Inbreeding effects may be more severe in non-domesticated
animals which have not undergone the intense selection and bottlenecks
which occur during the domestication process, making extrapolations to
the wild difficult.

The $N_e = 500$ estimate may not be useful for stress response traits
because it assumes that levels of genetic variance are only influenced by
a balance between mutation and drift or a balance between stabilizing
selection and mutation. As we have discussed in Section 5.7, genetic

variance in stress response traits may be influenced by other factors, such as selection in heterogeneous environments and trade-offs between life-history traits. Maintaining genetic variation via these mechanisms will depend on the environment as well as on the population size. In addition, levels of genetic variance can change during the course of selection. Franklin (1980) argued that selection does not generally influence genetic variance, and cited artificial selection experiments which produced a continuous selection response. However experiments with morphological characters commonly used in quantitative genetics, such as bristle number, do not always produce continuous responses. Traits under selection often show periods of accelerated responses reflecting changes in the realized heritability (Lee and Parsons 1968). The heritability of traits involved in stress resistance can also change during the course of selection: for example, heritabilities for knockdown resistance to ethanol changed within 15 generations of selection in *Drosophila pseudoobscura* and *D. simulans* (Hoffmann and Cohan 1987; Cohan and Hoffmann 1989). More spectacular examples are provided by some of the experiments with stress evasion traits discussed in Section 4.5, indicating that populations may become fixed for alleles responsible for genetic variation in dormancy in just a few generations. Directional selection or selection in heterogeneous environments can therefore have marked effects on the levels of genetic variation for stress response traits within a population, even in large populations (Parsons 1989*b*).

The $N_e = 500$ estimate depends on traits being controlled by a large number of loci. The impact of mutations on phenotypic variance in traits will be reduced when they are only affected by a few loci, so that a larger population size will be required to counter the effects of drift. For example, Lande and Barrowclough (1987) suggested that more than a quarter of a million individuals will be required to maintain high levels of heterozygosity at an individual locus with selectively neutral alleles. Maintaining high levels of genetic variance by a balance between stabilizing selection and mutation also depends on a large number of loci, and Turelli (1984) has argued that little variance will be maintained when the mutation rates of characters are low, which is likely when only a few loci contribute to a trait with typical mutation rates of less than 10^{-4}. Responses to environmental stress may involve genetic changes in traits with a relatively simple genetic basis (Section 5.6), particularly in the case of traits associated with stress evasion and resistance to toxic chemicals. The $N_e = 500$ argument may therefore not be applicable to the maintenance of genetic variation in many stress response traits.

Another problem discussed extensively in conservation genetics is the impact of bottlenecks (sharp decreases in population size) on genetic variation. Such events may not reduce levels of heterozygosity much if a population expands rapidly directly after the bottleneck (Nei *et al.* 1975). In *Euphydryas editha bayensis*, for example, allozyme frequencies and heterozygosity remained relatively stable despite wide-ranging fluctuations of population size (Ehrlich 1983); however while overall levels of genetic variation may be little affected by bottlenecks, there is a high probability that rare alleles will be lost (Nei *et al.* 1975). This may be particularly important for stress response traits since adaptation to some environmental stresses, including resistance to insecticides and heavy metals, is likely to involve rare alleles with major effects. Although we do not have much information on the frequency of alleles contributing to variation in resistance to climatic stresses, the trade-off between performance under stressful and favourable conditions (Chapter 7) suggests that these alleles will generally be rare in populations not normally exposed to stressful conditions. Resistance alleles may therefore be particularly susceptible to loss by bottlenecks in populations originating from favourable environments.

Recovery of the genetic variance of a trait following a bottleneck will depend on the number of loci controlling that trait. Intermediate levels of heritability will be restored through mutation in a few hundred to a few thousand generations when a large number of loci are involved, but traits controlled by a few loci will require recovery times that are several orders of magnitude greater, even when a population expands to a large size (Lande and Barrowclough 1987). Genetic variation for some stress response traits may therefore take a long time to recover once alleles are lost. These long time scales are unrealistic for the conservation of a population facing an increasing rate of environmental change in the next century.

The above considerations indicate the importance of distinguishing the conservation of rare genes from that of common genes, and genetic variance in traits controlled by a large number of loci from genetic variance in traits controlled by only a few loci. At present, we have too little information on genetic variation in stress resistance traits to determine the usefulness of the viable population size concept and allozyme heterozygosity levels in the conservation of genetic variance in these traits.

8.2.2 A directed approach for stress response traits?

The 'greenhouse effect' scenario emphasizes the need to consider traits that may determine future survival under conditions of rapid climatic change. This means that the maximum potential of organisms to respond to directional selection under stressful conditions needs to be preserved.

Conservation might, therefore, be a matter of preserving genetic variation in specific traits rather than concentrating solely upon maximizing the genetic variation in a population. This directed approach has the aim of maximizing the ability of a population to respond to specified environmental factors by directional selection, while an undirected approach assumes that such factors cannot be specified. We now turn to some considerations that may help to maintain high levels of genetic variation for responses to rapid environmental changes.

Substantial genetic variation for responses to climatic changes, pollutants and other stresses appears the norm in most populations (Chapter 4). An exception may be resistance to high temperature stress: this has shown little response to artificial selection in an experiment with *D. melanogaster* (Morrison and Milkman 1978) and only a slow response in wasps (White *et al.* 1970). However increases in high temperature resistance may occur in some populations via stress proteins, body colour changes and a reduced metabolic rate (Sections 3.2, 4.1 and 6.2), and evolutionary responses to temperature change are indicated by population divergence for the kinetic parameters of enzymes (Section 3.1). Nevertheless, genetic changes in responses to temperature stress may be smaller than for other stresses such as desiccation and pollutants, particularly in populations from marginal areas characterized by periods of exposure to high temperature stress.

Predicting the ability of populations to respond to an environmental stress will require the characterization of genetic variance within populations, and cannot therefore be made on the basis of the stress response of a few individuals. Furthermore, the presence of additive genetic variance for responses to a stress within a population does not mean that a population will successfully adapt to that stress. Evolutionary changes can be limited by other factors, as was discussed with respect to populations at species borders. These populations are marginal because they occupy environments that are sub-optimal and where abundance in low. Conserved populations will become marginal when reserves deteriorate or when environmental change cannot be countered by moving populations to favourable habitats. The hypotheses listed in Table 8.1 are therefore relevant to species conservation and the discussion of these hypotheses highlights the fact that we are largely ignorant of the genetic factors which limit the ability of species to adapt to increasingly stressful conditions. This applies to rapid environmental change as a consequence of human activities as well as more gradual environmental change. As already discussed, one of the few cases where some information on genetic factors limiting responses to environmental change is available is a response to heavy metal pollutants derived from mining activity.

Predicting population responses to stresses resulting from human activities will be complicated because of interactions between environmental factors. While the major predicted direct effect of the greenhouse scenario is an increase in global temperature, this stressor cannot be considered in isolation. In natural situations, temperature alterations and increasing chemical pollution are likely to occur together and it is, therefore, important to consider the combined effects of pollutants and other stresses on organisms. SO_2 and NO_2 have been shown to act synergistically on grasses (Ashenden and Mansfield 1978): this amplification of the separate effects of these chemicals on plant growth may explain the rapid development of SO_2 resistance in urban populations of plants, and the appearance of resistance to acute doses of SO_2 at rarely encountered levels (Mansfield and Freer-Smith 1981). There are also interactions between SO_2 and cadmium in the garden pea and cress (Czuba and Ormrod 1981; Hutchinson 1984): injury to cress occurs at lower concentrations of cadmium and O_3 when plants are exposed to these stresses together rather than separately. When temperature becomes stressful, interactions with pollutants may cause some areas to be severely affected (Vernberg and Vernberg 1974). For example, acute exposure to sublethal concentrations of cadmium reduced the thermal resistance of the small coastal North American fish, *Fundulus heteroclitus* (McKenney and Dean 1974); Barnes *et al.* (1988) found that exposure of the pea, *Pisum sativum*, to O_3 increased freezing injury. There is also evidence that forest decline occurs because of interactions between stresses: there are several primary factors and a multitude of secondary factors involved in such decline, and most of the six hypotheses concerning the primary factors involve stress interactions (Hinrichsen 1986). One hypothesis is that air pollutants and the deposition of nutrients and toxins interact to produce a poorer energy status, increasing susceptibility to other stresses. In another hypothesis involving soil acidification, nutrients are lost and toxic aluminium is released because of acidification; the aluminium damages roots and increases susceptibility to drought stress.

There is a lack of information on the evolution of resistance to multiple environmental stresses. We have seen in previous chapters that many stresses have a metabolic cost and this should tend to make the effects of different stresses cumulative, so that increased resistance to one stress will also increase resistance to another stress when both stresses are experienced by the same individual. In addition, the possibility that responses to diverse stresses involve common mechanisms (Chapter 6) will facilitate the evolution of increased resistance to multiple stresses in a population, even when multiple stresses are not experienced by the same individual. Experiments on genetic changes in response to multiple stresses, and particularly on the interaction of temperature stress with an array of more specific stresses, are required.

Although we cannot yet predict the ability of species to adapt to multiple stresses, management strategies that may be useful in ensuring the conservation of genetic variation can be suggested. A starting point is the realization that populations of plants and animals differ in mean levels of stress resistance, as discussed in Chapter 4; this helps in deciding which populations to conserve. Populations from ecological margins are most likely to have been selected for resistance to environmental stresses and will therefore contain genotypes useful in countering future stressful periods. The climatic matching approach described in Section 2.3 indicates the usefulness of considering genetic variation in responses to climatic factors when making predictions about the likely performance of populations in new environments; Burt et al. (1976) were able to predict the agronomic performance of strains of a pasture legume on the basis of their climate of origin. The possible existence of generalized stress resistance (Chapter 6) means that genotypes selected under extremes of one type of environmental stress may also exhibit increased resistance to a range of other environmental stresses. This strategy may therefore be useful even when conserved populations encounter future stresses that are not directly important at present species margins.

In contrast, the metabolic and structural costs of increased stress resistance (Chapter 7) suggest that alleles increasing stress resistance will be at lower frequencies in populations from favourable environments. The conservation of such populations may be less useful in a strategy that aims to maintain high levels of stress resistance to ensure survival during future environmental change. Selection of populations for conservation on the basis of stress resistance may, therefore, conflict with their selection on the basis of overall levels of genetic variation. Populations of some species from unfavourable environments may have lower levels of allozyme variation because of a reduced population size, while populations from favourable regions where species abundance is high and population sizes are large will tend to show the greatest allozyme variation.

An additional complication when deciding which populations to conserve is that populations may differ in responses to directional selection for increased stress resistance, even when they have similar levels of resistance prior to selection. This was shown for ethanol knockdown resistance in *Drosophila melanogaster* and *D. pseudoobscura* (Cohan and Hoffmann 1986; Hoffmann and Cohan 1987). Genetic differences between populations under the same selection pressure may develop because of differences in the heritability of the selected trait, or because populations respond by different underlying mechanisms. Ideally, all populations which are likely to differ in their response to selection should be conserved if the potential of a species to survive periods of environmental stress is to be maximized (Cohan and Hoffmann 1989). The ability of a species to adapt to a stress will be enhanced if populations which respond by different mechanisms are combined. Potential responses to directional selection could be identified by examining heritabilities and genetic correlations among traits within populations, determined from experiments over two generations comparing parents with their offspring or with a half-sib design (see Section 6.5). Phenotypic correlations may also be useful because they reflect genetic correlations in many situations, providing sample sizes are large (Cheverud 1988).

This emphasis upon the phenotypic level contrasts with approaches based upon the use of allozyme markers in the examination of population differences for conservation purposes (Parsons 1989*b*). Allozymes are of more limited usefulness: similar gene frequencies at a sample of loci do not indicate that populations will respond similarly to the same environmental stress. For example, *D. melanogaster* and *D. pseudoobscura* populations show only minor allozyme differences presumably because of the high vagility of these species, but the populations of these species still responded differently to selection (Cohan and Hoffmann 1989). This may partly reflect the fact that a comparison of gene frequencies at a random sample of loci provides no information on interactions between genes important in a selection response. Genetic differences between populations may also be more difficult to detect with allozymes than with quantitative traits (Lewontin 1984; Hoffmann *et al.* 1984).

How should populations be managed if high levels of stress resistance and high levels of genetic variation in stress response traits are to be maintained? The benefits of exposing populations to natural selection for ensuring that favourable genes persist at high frequencies are often emphasized in the conservation literature (Frankel and Soulé 1981; Schonewald-Cox *et al.* 1983). As is evident from the discussion in Chapter 7, it becomes difficult to define which genotypes will be favoured by natural selection when environments include both stressful and benign conditions. Genotypes favoured under optimal conditions may not be favoured under stressful conditions because of trade-offs between resistance and optimal performance. The maintenance of high levels of stress resistance in a population may therefore require natural selection under stressful conditions. This problem is particularly relevant to zoo populations destined for eventual release in the wild; individuals from these populations will not normally be exposed to non-optimal levels of abiotic factors, and there is evidence for their lowered capacity to adapt to extreme temperatures, especially in captive animals housed in temperature-regulated facilities (Kohane and Parsons 1988).

One way of maintaining genetic variation for stress resistance traits is to incorporate heterogeneous conditions into management programs for natural populations. Such heterogeneity could take the form of spatial variation in habitat types rather than temporal variation, avoiding the danger of extinction if lethality levels are exceeded during stressful periods. Species should ideally be conserved in areas that include relatively unfavourable habitats, where species have low densities consequent upon physical stress. The emphasis should not be solely on areas supporting a high population density, with the aim of maximizing population size.

One management question that has been extensively considered in the conservation literature is whether species should be conserved in one large or several small areas (see Simberloff 1988). The answer to this question may depend partly on the nature of their habitat. Species may be found in small fragmented areas that differ markedly from adjacent areas, such as remnant patches of rainforest. These may be particularly prone to disturbance and the abrupt environmental changes at rainforest margins may limit the ability of a species to respond genetically to changing conditions, so that they are best conserved in large reserves which are less likely to be disturbed. On the other hand, the potential role of spatial heterogeneity in maintaining genetic variation in stress response traits (Section 5.7) suggests that it may be more beneficial to conserve less restricted species in areas encompassing a range of habitat types, including marginal conditions. This may require several reserves or a single population in a large area with heterogeneous conditions. Conserving a species in a series of semi-isolated populations has the added benefit that stress resistant genotypes at the edge of species distributions are less likely to be swamped by an influx of genotypes from more optimal environments where numbers are higher. This is analogous to the situation where range expansion via resistant genotypes adapted to marginal conditions is restricted by gene flow from a central population (Section 8.1). A series of subdivided populations should also include favourable habitats where selection for intraspecific competitive ability is expected to be more intense because marginal environments may select for reduced competitive ability.

Managing populations for the maintenance of genetic variation is facilitated by monitoring changes in levels of genetic variation, preferably at loci that are likely to be important in future responses to environmental change. While monitoring for both electrophoretic variation and quantitative genetic variation in morphological traits has been recommended (Soulé 1987; Lande and Barrowclough 1987), these often make a minor contribution to changes in stress responses. Variation in most body measurements tends to reflect variation in overall body size, which may also make only a minor contribution to changes in stress resistance (e.g. Hoffmann and Parsons 1989a).

Although genetic variation in stress response traits should ideally be monitored directly in populations (Chapter 4), measures of stress resistance often involve mortality after exposure to an environmental stress. This difficulty could be overcome by the development of non-destructive methods of measuring stress resistance, which will be facilitated if resistance to different stresses involves similar mechanisms. Physiological and biochemical parameters such as metabolic rate, AECs and heat-shock proteins could provide useful ways of measuring genetic variation in resistance to many stresses; they are influenced by a range of stressful conditions (Section 1.3), and measures of developmental instability such as asymmetry in morphological traits could used in a similar way. The selection experiment of Frisch (1981) on beef cattle (Section 6.2) illustrates the ways in which stress resistance of even large mammals can be monitored in a non-destructive manner. Another potential problem is that heritabilities for stress resistance traits depend on stress levels (Section 5.4) and that alleles increasing fitness under stressful conditions will not be beneficial under favourable conditions (Chapter 7); conditions of severe stress relevant to anticipated environmental changes will need to be simulated when measurements of genetic parameters are made.

The focus of this discussion has been on stress responses, but the effects of climatic changes may also be mediated by biotic factors such as predation and disease. The conservation of genetic variation in traits relevant to these factors may be equally important. Failure to disperse and/or re-establish in other communities will also contribute to distribution changes and extinctions, emphasizing the potential significance of genetic variation in dispersal ability.

8.2.3 Concluding remarks

The above discussion highlights factors of potential importance in managing populations for survival under changing climatic conditions. The extent to which natural populations can respond to environmental stress is unclear, as are processes at species borders. We know little about factors that limit the ability of species to respond to present environmental conditions, let alone future changes in these conditions. There is a paucity of information on genetic variation in stress response traits under natural conditions, particularly from marginal populations. In addition to levels of genetic variation, evolutionary responses to changing environments will depend on population structure, inter-actions between different traits, and the history of selection in a popu-lation.

Genetic responses will be inadequate if environmental change is too rapid or too drastic, and failure to adapt to increases in environmental stress will result in species distribution changes. This means that the predicted greenhouse effect will lead to geographic range expansion by certain stress-resistant species into habitats such as rainforests if resist-ance levels of populations from these habitats are exceeded. For such assessments, species of the genus *Drosophila* have been shown to be important indicators of habitat change (Parsons 1989*b*). Predicting the ability of a species to evolve in response to rapid environmental change is a major task for evolutionary biology: the factors limiting species distributions, and the characteristics of populations near the margins, compared with more centrally located populations, should be given a high priority for future research.

8.3 Summary

Responses to environmental stress are important at species borders, and genetic variation in stress response traits underlies many hypotheses about limits to range expansion. There are insufficient data to evaluate the various hypotheses, although some direct and indirect evidence suggests that genetic variation in marginal environments may be limiting in some cases because of physiological constraints.

Stress response traits are important determinants of the future survival of species under changing global climatic conditions and localized pollution. In contrast, conservation genetics has focused on the maintenance of maximum genetic variation rather than a phenotypic and genetic analysis of stress response traits. Published recommendations concerning population size and effects of bottlenecks appear only partly relevant to many stress response traits because of the likely importance of rare alleles and the small number of loci influencing variance in some of these traits.

Genetic variation in stress response traits and high levels of stress resistance may be conserved if populations are established with an emphasis on individuals from marginal habitats, and if populations are maintained in reserves that include areas with marginal conditions. A reassessment of conservation strategies and procedures is necessary under a scenario of rapidly changing environmental conditions.

References

Abel, G.H. (1969). Inheritance of the capacity for chloride inclusion and chloride exclusion of soybeans. *Crop Science* **9**, 697–8.

Agnew, A.D.Q. (1968). Variation and selection in an isolated series of populations of *Lysimachia volkensii* Engl. *Evolution*, **22**, 228–36.

Alahiotis, S.N. (1982). Adaptation of *Drosophila* enzymes to temperature. IV. Natural selection at the alcohol dehydrogenase locus. *Genetics*, **59**, 81–7.

Alahiotis, S.N., and Stephanou, G. (1982). Temperature adaptation of *Drosophila* populations. The heat-shock proteins system. *Comparative Biochemistry and Physiology*, **73(B)**, 529–33.

Allen, F.L., Comstock, R.E., and Rasmusson, D.C. (1978). Optimal environments for yield testing. *Crop Science*, **18**, 747–51.

Alvarez, L.W., Alvarez, W., Asaro, F., and Michel, H.V. (1980). Extraterrestrial cause for the Cretaceous and Tertiary extinction. *Science*, **208**, 1095–1108.

Ananthan, J., Goldberg, A.L., and Voellmy, R. (1986). Abnormal proteins serve as eukaryotic stress signals and trigger the activation of heat shock genes. *Science*, **232**, 522–4.

Anderson, B.A., and Karasov, W.H. (1981). Contrasts in energy intake and expenditure in sit-and-wait and widely foraging lizards. *Oecologia*, **49**, 67–72.

Anderson, W.W., Arnold, J., Sammons, S.A., and Yardley, D.G. (1986). Frequency-dependent viabilities of *Drosophila pseudoobscura* karyotypes. *Heredity*, **56**, 7–17.

Andrewartha, H.G., and Birch, L.C. (1954). *The distribution and abundance of animals.* University of Chicago Press, Chicago.

Antlfinger, A.E. (1981). The genetic basis of microdifferentiation in natural and experimental populations of *Borrichia frutescens* in relation to salinity. *Evolution*, **35**, 1056–63.

Antonovics, J. (1976). The nature of limits to natural selection. *Annals of the Missouri Botanical Garden*, **63**, 224–47.

Antonovics, J., and Primack, R.B. (1982). Experimental ecological genetics in *Plantago*. VI. The demography of seedling transplants of *P. lanceolata*. *Journal of Ecology*, **70**, 55–75.

Antonovics, J., Bradshaw, A.D., and Turner, R.G. (1971). Heavy metal tolerance in plants. *Advances in Ecological Research*, **7**, 1–85.

Arking, R., and Clare, M.J. (1986). Genetics of aging: effective selection for increased longevity in *Drosophila*. In *Insect aging* (ed. K.-G. Collatz and R.S. Sohal), pp. 217–36. Springer-Verlag, Berlin.

Arking, R., Buck, S., Wells, R.A., and Pretzlaff, R. (1988). Metabolic rates in genetically based long lived strains of *Drosophila*. *Experimental Gerontology*, **23**, 59–76.

Arnason, E., and Grant, P.R. (1976). Climatic selection in *Cepaea hortensis* at the northern limit of its range in Iceland. *Evolution*, **30**, 499–508.

Arthur, A.E., Gale, J.S., and Lawrence, M.J. (1973). Variation in wild populations of *Papaver dubium*. VII. Generation time. *Heredity*, **30**, 1989–97.

Ashenden, T.W., and Mansfield, T.A. (1978). Extreme pollution sensitivity of grasses when SO_2 and NO_2 are present in the atmosphere together. *Nature*, **273**, 142–3.

Ashenden, T.W., Stewart, W.S., and Williams, W. (1975). Growth responses of sand dune populations of *Dactylis glomerata* L. to different level of water stress. *Journal of Ecology*, **63**, 97–107.

Ashraf, M., McNeilly, T., and Bradshaw, A.D. (1987). Selection and heritability of tolerance to sodium chloride in four forage species. *Crop Science*, **227**, 232–234.

Aslanyan, M.M., Smirnova, V.A., and Prokof'eva, L.P. (1987). The nature of induced recombination in males of the line mei—9^{LI} of *Drosophila melanogaster* under the influence of space flight factors. *Genetika*, **23**, 193–5.

Atkinson, D.E. (1977). *Cellular energy metabolism and its regulation.* Academic Press, New York.

Attfield, P.V. (1987). Trehalose accumulates in *Saccharomyces cerevisiae* during exposure to agents that induce heat shock response. *FEBS Letters*, **255**, 259–63.

Ayres, P.G. (1984). The interaction between environmental stress injury and biotic disease physiology. *Annual Review of Phytopathology*, **22**, 53–75.

Baker, A.J.M., Grant, C.J., Martin, M.H., Shaw, S.C., and Whitebrook, J. (1986). Induction and loss of cadmium tolerance in *Holcus lanatus* L. and other grasses. *New Phytologist*, **102**, 575–87.

Baker, R.L., and Cockrem, F.R.M. (1970). Selection for body weight in the mouse at three temperatures and the correlated response in tail length. *Genetics*, **63**, 505–23.

Baldwin, J.D, and Dingle, H. (1986). Geographic variation in the effects of temperature on life-history traits in the large milkweed bug *Oncopeltus fasciatus*. *Oecologia*, **69**, 64–71.

Bantock, C.R. (1980). Variation in the distribution and fitness of the brown morph of *Cepaea nemoralis* (L.). *Biological Journal of the Linnean Society*, **13**, 47–64.

Bantock, C.R., and Price, D.J. (1975). Marginal population of *Cepaea nemoralis* (L.) on the Brendon Hills, England. I. Ecology and ecogenetics. *Evolution*, **29**, 267–77.

Barker, J.S.F. (1983). Interspecific competition. In *The genetics and biology of Drosophila, Volume 3c* (eds. M. Ashburner, H.L. Carson and J.N. Thompson), pp. 284–341. Academic Press, London.

Barlow, R. (1981). Experimental evidence for interactions between heterosis and environment in animals. *Animal Breeding Abstracts*, **49**, 715–37.

Barnes, J.D., Reiling, K., Davison, A.W., and Renner, C.J. (1988). Interaction between ozone and winter stress. *Environmental Pollution*, **53**, 235–54.

Barrow, N.J. (1977). Phosphorus uptake and utilization by tree seedlings. *Australian Journal of Botany*, **25**, 571–84.

Barton, N.H., and Turelli, M. (1990). Evolutionary quantitative genetics: How little do we know?. *Annual Review of Genetics*, **23**, 337–70.

Bateman, K.G. (1959). The genetic assimilation of four venation phenocopies. *Journal of Genetics*, **56**, 443–74.

Battaglia, B. (1967). Genetic aspects of benthic ecology in brackish waters. In *Estuaries* (ed. G.H. Lauft), pp. 574–7. American Association for the Advancement of Science, Washington.

Bayne, B.L. (1975). Aspects of physiological condition in *Mytilus edulis* L. with respect to the effects of oxygen tension and salinity. *Proceedings of the Ninth European Marine Biology Symposium*, 213–38.

Bazzaz, F.A., and Carlson, R.W. (1982). Photosynthetic acclimation to variability in the light environment of early and late successional plants. *Oecologia*, **54**, 313–6.

Beardmore, J.A. (1960). Developmental stability in constant and fluctuating temperatures. *Heredity*, **14**, 411–22.

Beaumont, A.R., Beveridge, C.M., Barnet, E.A., Budd, M.D., and Smyth-Chamosa, M. (1988). Genetic studies of laboratory reared *Mytilus edulis*. I. Genotype specific selection in relation to salinity. *Heredity*, **61**, 389–400.

Beaumont, M.A. (1988). Stabilizing selection and metabolism. *Heredity*, **61**, 433–8.

Beis, I., and Newholme, E.A. (1975). The contents of adenine nucleotides, phosphagens and some glycolytic intermediates in resting muscles from vertebrates and invertebrates. *Biochemical Journal*, **152**, 23–32.

Bell, G. (1984a). Measuring the cost of reproduction. I. The correlation structure of the life table of a plankton rotifer. *Evolution*, **38**, 300–13.

Bell, G. (1984b). Measuring the cost of reproduction. II. The correlation structure of the life tables of five fresh water invertebrates. *Evolution*, **38**, 314–26.

Belyaev, D.K., and Borodin, P.M. (1982). The influence of stress on variation and its role in evolution. *Biologisches Zentralblatt*, **100**, 705–14.

Berry, J.A. (1975). Adaptation of photosynthetic processes to stress. *Science*, **118**, 644–50.

Berry, J.A., and Bjorkman, O. (1980). Photosynthetic response and adaptation to temperature in higher plants. *Annual Review of Plant Physiology*, **31**, 491–543.

Berven, K.A., and Gill, G.E. (1983). Interpreting geographic variation in life history traits. *American Zoologist*, **23**, 85–97.

Beversdorf, W.D., Hume, D.J., and Donnelly-Vanderloo, M.J. (1988). Agronomic performance of Triazine-resistant and susceptible reciprocal spring Canola hybrids. *Crop Science*, **28**, 932–4.

Billings, W.D., Godfrey, P.J., Chabot, B.F., and Bourque, D.P. (1971). Metabolic acclimation to temperature in arctic and alpine ecotype of *Oxyria digyna*. *Arctic and Alpine Research*, **3**, 277–89.

Bjorkman, O. (1966). Comparative studies of photosynthesis and respiration in ecological races. *Brittonia*, **18**, 214–24.

Blackstock, J. (1984). Biochemical metabolic regulatory responses of marine invertebrates to natural environmental change and marine pollution. *Oceanography and Marine Biology Annual Reviews*, **22**, 263–313.

Blum, A. (1988). *Plant breeding for stress environments*. CRC Press, Boca Raton.

Boag, P.T., and Grant, P.R. (1978). Heritability of external morphology in Darwin's finches. *Nature*, **274**, 793–4.

Boag, P.T., and Grant, P.R. (1981). Intense natural selection in a population of Darwin's finches (Geospizinae) in the Galapagos. *Science*, 214, 82–5.

Bohren, B.B., Carson, J.R., and Rogler, J.C. (1981). Response to selection at two temperatures for fast and slow growth from five to nine weeks of age in poultry. *Genetics*, 97, 443–56.

Borodin, P.M. (1987). Stress and genetic variability. *Genetika*, 23, 1003–10.

Boulétreau, J. (1978). Ovarian activity and reproductive potential in a natural population of *Drosophila melanogaster*. *Oecologia*, 33, 319–42.

Bouletreau-Merle, J., Fouillet, P., and Terrier, O. (1987). Seasonal variations and balanced polymorphisms in the reproductive potential of temperate *D. melanogaster* populations. *Entomologia experimentalis et applicata*, 43, 39–48.

Bowen, S.J., and Washburn, K.W. (1984). Genetics of heat tolerance in Japanese quail. *Poultry Science*, 63, 430–5.

Boyce, M.S., and Perrins, C.M. (1987). Optimizing great tit clutch size in a fluctuating environment. *Ecology*, 58, 142–53.

Boyer, J.S. (1982). Plant productivity and environment. *Science*, 218, 443–8.

Bradley, B.P. (1978). Genetic and physiological adaptation of the copepod *Eurytemora affinis* to seasonal temperature. *Genetics*, 90, 193–205.

Bradley, B.P. (1981). Models for physiological and genetic adaptation to variable environments. In *Insect life history patterns: habitats and genetic variation* (eds. R.F. Denno and H. Dingle), pp. 33–50. Springer-Verlag, Berlin.

Bradshaw, A.D. (1965). Evolutionary significance of phenotypic plasticity in plants. *Advances in Genetics*, 13, 115–53.

Bradshaw, A.D. (1984). Adaptation of plants to soils containing toxic metals—a test for conceit. In *Origin and development of adaptation* (eds. D. Evered, and G.M. Collins), pp. 4–19. CIBA Foundation Symposium 102. Pitman, London.

Bradshaw, A.D., Chadwick, M.J., Jowett, D., and Snaydon, R.W. (1965). Experimental investigations into the mineral nutrition of several grass species. *Journal of Ecology*, 52, 665–76.

Branch, G.M., Borchers, P., Brown, C.R., and Donnelly, D. (1988). Temperature and food as factors influencing oxygen consumption of intertidal organisms, particularly limpets. *American Zoologist*, 28, 137–46.

Brand, M.D., and Murphy, M.P. (1987). Control of electron flux through the respiratory chain in mitochondria and cells. *Biological Reviews*, 62, 141–93.

Brett, J.R. (1970). Temperature. Animals. Fishes. In *Marine Ecology*, Vol. I, Part 1 (ed. O. Kinne), 515–60. Wiley, London.

Broadhurst, P.L. (1979). The experimental approach to behavioural evolution. In *Theoretical advances in behaviour genetics* (eds. J.R. Royce and L.P. Moss), pp. 43–95. Sijthoff and Noordhoff, Alphen aan de Rijn.

Brown, A.H.D., Marshall, D.R., and Munday, J. (1976). Adaptedness of variants at an alcohol dehydrogenase locus in *Bromus mollis* L. (soft brome-grass). *Australian Journal of Biological Sciences*, 29, 389–96.

Browne, R.A., Davis, L.E., and Sallee, S.E. (1988). Effects of temperature and relative fitness of sexual and asexual brine shrimp *Artemia*. *Journal of Experimental Marine Biology and Ecology*, 124, 1–20.

Brubaker, L.B. (1986). Responses of tree populations to climatic change. *Vegetatio*, 67, 119–30.

Bruckner, P.L., and Frohberg, R.C. (1987). Stress tolerance and adaptation in spring wheat. *Crop Science*, **27**, 31–6.

Brussard, P.F. (1984). Geographic patterns and environmental gradients: the central-marginal model in *Drosophila* revisited. *Annual Review of Ecology and Systematics*, **15**, 25–64.

Buddenhagen, I.W. (1983). Breeding strategies for stress and disease resistance in developing countries. *Annual Review of Phytopathology*, **21**, 385–409.

Bullock, T.H. (1955). Compensation for temperature in the metabolism and activity of poikilotherms. *Biological Reviews*, **30**, 311–42.

Burt, R.L., Reid, R., and Williams, W.T. (1976). Explanation for, and utilization of, collections of tropical pasture legumes. *Agro-Ecosystems*, **2**, 293–307.

Burton, R.S., and Feldman, M.W. (1983). Physiological effects of an allozyme polymorphism: glutamate–pyruvate transaminase and response to hyperosmotic stress in the copepod *Tigriopus californicus*. *Biochemical Genetics*, **21**, 239–51.

Burton, V., Mitchell, H.K., Young, P., and Petersen, N.S. (1988). Heat shock protection against cold stress of *Drosophila melanogaster*. *Molecular and Cellular Biology*, **8**, 3550–2.

Busby, J.R. (1986*a*). *Bioclimate prediction system*. User's Manual, Bureau of Fauna and Flora, Canberra.

Busby, J.R. (1986*b*). A biogeoclimatic analysis of *Nothofagus cunninghamii* (Hook.) Oerstn. in southeastern Australia. *Australian Journal of Ecology*, **11**, 1–7.

Calabrese, E.J., McCarthy, M.E., and Kenyon, E. (1987). The occurrence of chemically induced hormesis. *Health Physics*, **52**, 531–41.

Calvert, W.H., Zuchowski, W., and Brower, L.P. (1983). The effect of rain, snow and freezing temperatures on overwintering Monarch butterflies in Mexico. *Biotropica*, **15**, 42–7.

Capinera, J.L. (1979). Qualitative variation in plants and insects: effect of propagule size on ecological plasticity. *American Naturalist*, **114**, 350–61.

Carmi-Winkler, N., Degen, A.A., and Pinshow, B. (1987). Seasonal time-energy budgets of free-living chukars in the Negev desert. *Condor*, **89**, 594–601.

Carne, P.B. (1962). The characteristics and behaviour of the saw-fly *Perga affinis affinis* (Hymenoptera). *Australian Journal of Zoology*, **10**, 1–34.

Carson, H.L. (1958). Response to selection under different conditions of recombination in *Drosophila*. *Cold Spring Harbor Symposia on Quantitative Biology*, **23**, 291–306.

Carson, H.L. (1959). Genetic conditions that promote or retard the formation of species. *Cold Spring Harbor Symposia on Quantitative Biology*, **24**, 87–103.

Carvalho, G.R. (1987). The clonal ecology of *Daphnia magna* (Crustacea: Cladocera). II. Thermal differentiation among seasonal clones. *Journal of Animal Ecology*, **56**, 469–78

Carvalho, G.R., and Crisp, D.J. (1987). The clonal ecology of *Daphnia magna* (Crustacea: Cladocera). I. Temporal changes in the clonal structure of a natural population. *Journal of Animal Ecology*, **56**, 453–68.

Caughley, G., Grigg, G.C., and Smith, L. (1985). The effect of drought on kangaroo populations. *Journal of Wildlife Management*, **49**, 679–85.

Caughley, G., Short, J., Grigg, G.C., and Nix, H. (1987). Kangaroos and climate: an analysis of distribution. *Journal of Animal Ecology*, **56**, 751–61.

Ceccarelli, S. (1987). Yield potential and drought tolerance of segregating populations of barley in contrasting environments. *Euphytica*, **36**, 265–73.

Ceccarelli, S. (1989). Wide adaptation: How wide? *Euphytica*, **40**, 197–205.

Chambers, G.K. (1988). The *Drosophila* alcohol dehydrogenase gene-enzyme system. *Advances in Genetics*, **25**, 39–107

Chapin, F.S. (1980). The mineral nutrition of wild plants. *Annual Review of Ecology and Systematics*, **11**, 233–60.

Chapin, F.S., Follett, J.M., and O'Conner, K.F. (1982). Growth, phosphate absorption, and phosphorus chemical fractions in two *Chionochloa* species. *Journal of Ecology*, **70**, 305–21.

Chappell, M.A., and Bartholomew, G.A. (1981). Standard operative temperatures and thermal energetics of the antelope ground squirrel *Ammospermophilus leucurus*. *Physiological Zoology*, **54**, 81–93.

Chappell, M.A., and Snyder, L.R.G. (1984). Biochemical and physiological correlates of deer mouse alpha-chain hemoglobin polymorphisms. *Proceedings of the National Academy of Sciences USA*, **81**, 5484–8.

Charlesworth, B. (1976). Recombination modification in a fluctuating environment. *Genetics*, **83**, 181–95.

Charlesworth, B. (1979). Evidence against Fisher's theory of dominance. *Nature*, **278**, 848–9.

Charlesworth, B., and Charlesworth, D. (1985). Genetic variation in recombination in *Drosophila*. I. Responses to selection and preliminary genetic analysis. *Heredity*, **54**, 71–83.

Charnov, E.L., and Schaffer, W.M. (1973). Life history consequences of natural selection: Cole's result revisited. *American Naturalist*, **107**, 791–3.

Cheeseman, J.M. (1988). Mechanisms of salinity tolerance in plants. *Plant Physiology*, **87**, 577–80.

Cheverud, J.M. (1984). Quantitative genetics and developmental constraints on evolution by selection. *Journal of Theoretical Biology*, **110**, 155–71.

Cheverud, J.M. (1988). A comparison of genetic and phenotypic correlations. *Evolution*, **42**, 958–68.

Cicerone, R.J. (1987). Changes in stratospheric oxygen. *Science*, **237**, 35–42.

Clare, M.J., and Luckinbill, L.S. (1985). The effects of gene–environment interaction on the expression of longevity. *Heredity*, **55**, 19–25.

Clark, A.G. (1987). Senescence and the genetic-correlation hang-up. *American Naturalist*, **129**, 932–40.

Clarke, B. (1979). The evolution of genetic diversity. *Proceedings of the Royal Society of London Series B*, **205**, 453–74.

Clausen, J., Keck, D.D., and Heisey, W.M. (1940). Experimental studies on the nature of species. *Carnegie Institute of Washington Publication*, **520**, 1–452.

Clausen, J., Keck, D.D., and Hiesey, W.M. (1948). Experimental studies on the nature of species. III. Environmental responses of climatic races of *Achillea*. *Carnegie Institute of Washington Publication 581*.

Clegg, M.T., and Allard, R.W. (1972). Patterns of genetic differentiation in the slender wild oat species, *Avena barbata*. *Proceedings of the National Academy of Sciences, USA*, **69**, 1820–4.

Clough, J.M., Alberte, R.S., and Teeri, J.A. (1979). Photosynthetic adaptation of *Solanum dulcamara* L. to sun and shade environments. *Plant Physiology*, **64**, 25–30.

Cluster, P.D., Marinkovic, C., Allard, R.W., and Ayala, F.J. (1987). Correlations between development rates, enzyme activities, ribosomal DNA spacer-length phenotypes, and adaptation in *Drosophila melanogaster. Proceedings of the National Academy of Sciences, USA*, **84**, 610–4.

Coakley, S.M. (1988). Variation in climate and prediction of disease in plants. *Annual Review of Phytopathology*, **26**, 163–81.

Coffman, F.A. (1957). Cold-resistant oat varieties also resistant to heat. *Science*, **125**, 1298–9.

Cohan, F.M., and Hoffmann, A.A. (1986). Genetic divergence under uniform selection. II. Different responses to selection for knockdown resistance to ethanol among *Drosophila melanogaster* populations and their replicate lines. *Genetics*, **114**, 145–63.

Cohan, F.M., and Hoffmann, A.A. (1989). Uniform selection as a diversifying force in evolution: evidence from Drosophila. *American Naturalist*, **134**, 613–37.

Cohan, F.M., Hoffmann, A.A., and Gayley, T.W. (1989). A test of the role of epistasis in divergence under uniform selection. *Evolution*, **43**, 766–74.

Connell, J.H. (1983). On the prevalence and relative importance of inter-specific competition: evidence from field experiments. *American Naturalist*, **122**, 661–96.

Cook, R.E. (1979). Patterns of juvenile mortality and recruitment of plants. In *Topics in plant population biology* (eds. O.T. Solbrig, S. Jain, G.B. Johnson and P.H. Raven), pp. 207–31. Columbia University Press, New York.

Cowie, R.H., and Jones, J.S. (1985). Climatic selection on body colour in *Cepaea. Heredity*, **55**, 261–7.

Cox, E.C., and Gibson, T.C. (1974). Selection for high mutation rates in chemostats. *Genetics*, **77**, 169–84.

Coyne, J.A., and Beecham, E. (1987). Heritability of two morphological characters within and among natural populations of *Drosophila melanogaster. Genetics*, **117**, 727–37.

Coyne, J.A., Bundgaard, J., and Prout, T. (1983). Geographic variation of tolerance to environmental stress in *Drosophila pseudoobscura. American Naturalist*, **122**, 474–88.

Crawford, R.M.M. (1983). Root survival in flooded soils. In *Ecosystems of the World, Vol. 4A: Mires, swamp, bog, fen and moors* (ed. A.J.P. Gore), pp. 257–83. Elsevier, Amsterdam.

Crick, J.C., and Grime, J.P. (1987). Morphological plasticity and mineral nutrient capture in two herbaceous species of contrasted ecology. *New Phytologist*, **107**, 403–14.

Crow, J.F. (1957). Genetics of insect resistance to chemicals. *Annual Review of Entomology*, **2**, 227–46.

Crump, M.L. (1981). Variation in propagule size as a function of environmental uncertainty for tree frogs. *American Naturalist*, **117**, 724–37.

Cullis, C.A. (1987). The generation of somatic and heritable variation in response to stress. *American Naturalist*, **130**, 562–73.

Czuba, M., and Ormrod, D.P. (1981). Cadmium concentrations in cress shoots in relation to cadmium enhanced ozone phytotoxicity. *Environmental Pollution Series A, Ecology and Biology*, **25**, 67–76.

Daday, H., Binet, F.E., Grassia, A., and Peak, J.W. (1973). The effect of environment on heritability and predicted selection response in *Medicago sativa. Heredity*, **31**, 293–308.

Danzmann, R.G., and Ferguson, M.M. (1988). Developmental rates of heterozygous and homozygous rainbow trout reared at three temperatures. *Biochemical Genetics*, **26,** 53–67.

Darwin, C. (1859). *On the origin of species by mean of natural selection.* Murray, London.

Darwin, C., and Wallace, A.R. (1859). On the tendency of species to form varieties; and on the perpetuation of varieties and species by means of selection. *Journal of the Linnean Society of London (Zoology)*, **3**, 45–62.

David, J.R., Allemand, R., Van Herrewege, J., and Cohet, Y. (1983). Ecophysiology: abiotic factors. In *The genetics and biology of Drosophila*, Vol. 3d (eds. M. Ashburner, H.L. Carson, and J.N. Thompson), pp. 105–7. Academic Press, London.

Davies, M.S., and Snaydon, R.W. (1976). Rapid population differentiation in a mosaic environment. III. Measures of selection pressures. *Heredity*, **36**, 59–66.

Davis, R.W., Williams, T.M., Thomas, J.A., Kastelein, R.A., and Cornell, L.H. (1988). The effects of oil contamination and cleaning on sea otters (*Enhydra lutris*). II. Metabolism, thermoregulation, and behavior. *Canadian Journal of Zoology*, **66**, 2782–90.

Deery, B.J., and Parsons, P.A. (1972a). Variations in the resistance of natural populations of *Drosophila* to phenyl-thio-carbamide (P.T.C.). *Egyptian Journal of Genetics and Cytology*, **1**, 13–7.

Deery, B.J., and Parsons, P.A. (1972b). Ether resistance in *Drosophila melanogaster. Theoretical and Applied Genetics*, **42**, 208–14.

Den Boer, P.J. (1985). Fluctuations of density and survival of carabid populations. *Oecologia*, **67**, 322–30.

Denno, R.F., Olmstead, K.L., and McCloud, E.S. (1989). Reproductive cost of flight capability: a comparison of life history traits in wing dimorphic planthoppers. *Ecological Entomology*, **14**, 31–44.

Derr, J.A. (1980). The nature of variation in life history characters of *Dysdercus bimaculatus* (Heteroptera: Pyrrhocoridae), a colonizing species. *Evolution*, **34**, 548–57.

Diehl, W.J., Gaffney, P.M., and Koehn, R.K. (1986). Physiological and genetic aspects of growth in the mussel *Mytilus edulis*. I. Oxygen consumption, growth and weight loss. *Physiological Zoology*, **59**, 201–11.

DiMichele, L., and Powers, D.A. (1984). Developmental and oxygen consumption rate differences between lactate dehydrogenase-B genotypes of *Fundulus heteroclitus* and their effect on hatching time. *Physiological Zoology*, **57**, 52–6.

Din, Z.B., and Brooks, J.M. (1986). Use of adenylate energy charge as a physiological indicator in toxicity experiments. *Bulletin of Environmental Contamination and Toxicology*, **36**, 1–8.

Dingle, H. (1978). Migration and diapause in tropical, temperate, and island milkweed bugs. In *Evolution of insect migration and diapause* (ed. H. Dingle), pp. 254–76. Springer, New York.

Dingle, H., Evans, K.E., and Palmer, J.O. (1988). Responses to selection among life history traits in a nonmigratory population of milk weed bugs (*Oncopeltus fasciatus*). *Evolution*, **42**, 79–92.

Duke, E.J., and Glassman, E. (1968). Drug effects in *Drosophila*: streptomycin sensitive strains and fluorouracil resistant strains. *Nature*, **220**, 588–9.

Durrant, A. (1962). The environmental induction of heritable changes in Linum. *Heredity*, **17**, 27–61.

Durrant, A., and Timmis, J.N. (1973). Genetic control of environmentally induced changes in *Linum*. *Heredity*, **30**, 369–79.

Dykhuizen, D.E., and Hartl, D.L. (1980). Selective neutrality of 6PGD allozymes in *E. coli* and the effects of genetic background. *Genetics*, **96**, 801–17.

Dykhuizen, D.E., and Hartl, D.L. (1983). Functional effects of PGI allozymes in *Escherichia coli*. *Genetics*, **105**, 1–18.

Dykhuizen, D.E., Dean, A.M., and Hartl, D.L. (1987). Metabolic flux and fitness. *Genetics*, **115**, 25–31.

Eanes, W.F. (1984). Viability interactions, in vivo activity, and the G6PD polymorphism in *Drosophila melanogaster*. *Genetics*, **106**, 95–107.

Easton, D.P., Rutledge, P.S., and Spotila, J.R. (1987). Heat shock protein induction and induced thermal tolerance are independent in adult salamanders. *Journal of Experimental Zoology*, **241**, 263–7.

Eberhardt, S.A., and Russell, W.A. (1966). Stability parameters for comparing varieties. *Crop Science*, **6**, 36–40.

Eckstrand, I.A., and Richardson, R.H. (1980). Comparison of some water balance characteristics in several *Drosophila* species which differ in habitat. *Environmental Entomology*, **9**, 716–20.

Ehrlich, P.R. (1983). Genetics and the extinction of butterfly populations. In *Genetics and conservation* (eds. C.M. Schonewald-Cox, S.M. Chambers, B. MacBryde, and W.L. Thomas), pp. 152–63. Benjamin Cummings, London.

Ehrlich, P.R., and Raven, P.H. (1969). Differentiation of populations. *Science*, **165**, 1228–32.

Ehrlich, P.R., Breedlove, D.E., Brussard, P.R., and Sharp, M.A. (1972). Weather and the 'regulation' of subalpine populations. *Ecology*, **53**, 243–7.

Ehrlich, P.R., Murphy, D.D., Singer, M.C., Sherwood, C.B., White, R.R., and Brown, I.L. (1980). Extinction, reduction, stability and increase: the responses of checkerspot butterfly (*Euphydryas*) populations to the California drought. *Oecologia*, **46**, 101–5.

Emeka-Ejiofor, S.A.I., Curtis, C.F., and Davidson, G. (1983). Tests for effects of insecticide resistance genes in *Anopheles gambiae* on fitness in the absence of insecticides. *Entomologia Experimentalis et Applicata*, **34**, 163–8.

Endler, J.A. (1986). *Natural selection in the wild*. Princeton University Press, Princeton.

Etter, R.J. (1988). Physiological stress and color polymorphism in the intertidal snail *Nucella lapillus*. *Evolution*, **42**, 660–80.

Falconer, D.S. (1952). The problem of environment and selection. *American Naturalist*, **36**, 293–8.

Falconer, D.S. (1960). Selection of mice for growth on high and low planes of nutrition. *Genetical Research*, **1**, 91–113.

Falconer, D.S. (1981). *Introduction to quantitative genetics*, 2nd edn. Longman, London.

Falconer, D.S., and Latyszewski, M. (1952). The environment in relation to selection for size in mice. *Journal of Genetics*, **51**, 67–80.

Fatt, H.V., and Dougherty, F.C. (1963). Genetic control of differential heat tolerance in two strains of the nematode *Caenorhabditis elegans*. *Science*, **142**, 266–7.

Felsenstein, J. (1976). The theoretical population genetics of variable selection and migration. *Annual Review of Genetics*, **10**, 253–80.

Finch, S., and Collier, R.H. (1983). Emergence of flies from overwintering populations of cabbage root fly pupae. *Ecological Entomology*, **8**, 29–36.

Finch, V.A., and Western, D. (1977). Cattle colors in pastoral herds: natural selection or social preference? *Ecology*, **58**, 1384–92.

Finlay, K.W., and Wilkinson, G.N. (1963). The analysis of adaptation in a plant-breeding programme. *Australian Journal of Agricultural Research*, **14**, 742–54.

Finley, D., Ozkaynak, E., and Varshavsky, A. (1987). The yeast polyubiquitin gene is essential for high temperatures, starvation, and other stresses. *Cell*, **48**, 1035–46.

Fischer, R.A., and Wood, J.T. (1979). Drought resistance spring wheat cultivars. III. Yield associations with morpho-physiological traits. *Australian Journal of Agricultural Research*, **30**, 1001–20.

Fisher, R.A. (1930). *The genetical theory of natural selection*. Clarendon Press, Oxford.

Fleischer, R.C., and Johnston, R.F. (1982). Natural selection on body size and proportions in house sparrows. *Nature*, **298**, 747–9.

Flexon, P.B., and Rodell, C.F. (1982). Genetic recombination and directional selection for DDT resistance in *Drosophila melanogaster*. *Nature*, **298**, 672–4.

Forman, R.T.T. (1964). Growth under controlled conditions to explain the hierarchical distributions of a moss, *Tetraphis pellucida*. *Ecological Monographs*, **34**, 1–25.

Forsum, E., Hillman, P.E., and Nesheim, M.C. (1981). Effect of energy restriction on total heat production, basal metabolic rate, and specific dynamic action of food in rats. *Journal of Nutrition*, **111**, 1691–7.

Foy, C.D. (1984). Adaptation of plants to mineral stress in problem soils. In *Origins and development of adaptation* (Ciba Foundation Symposium, 102), pp. 20–39. Pitman Books, London.

Frankel, O.H., and Soulé, M.E. (1981). *Conservation and evolution*, Cambridge University Press, Cambridge.

Frankham, R. (1980). Origin of genetic variation in selection lines. In *Selection experiments in laboratory and domestic animals* (ed. A. Robertson), pp. 56–68. Commonwealth Agricultural Bureaux.

Frankham, R. (1988). Exchanges in the rDNA multigene family as a source of genetic variation. In *Proceedings of the second international conference on quantitative genetics* (eds. B.S. Weir, E.J. Eisen, M.M. Goodman, and G. Namkoong), pp. 236–46. Sinauer, Sunderland.

Franklin, I.R. (1980). Evolutionary changes in small populations. In *Conservation biology. An evolutionary–ecological perspective* (eds. M.E. Soulé, B.A. Wilcox), pp. 135–45. Sinauer, Sunderland.

Frisch, J.E. (1981). Changes occurring in cattle as a consequence of selection for growth rate in a stressful environment. *Journal of Agricultural Science*, **96**, 23–38.

Frisch, J.E., and Vercoe, J.E. (1978). Utilizing breed differences in growth of cattle in the tropics. *World Animal Review*, **25**, 8–12.

Frumkin, R., Pinshow, B., and Weinstein, Y. (1986). Metabolic heat production and evaporative heat loss in desert Phasianids: chukar and sand partridge. *Physiological Zoology*, **59**, 592–605.

Fuchs, A., de Ruig, S.P., van Tuyl, J.M., and de Vries, F.W. (1977). Resistance to triforine: A non existent problem? *Netherlands Journal of Plant Pathology*, **83 (suppl.)**, 189–205.

Futuyma, D. (1979). *Evolutionary biology*. Sinauer, Sunderland.

Gartside, D.W., and McNeilly, T. (1974). Genetic studies in heavy metal tolerant plants. Genetics of zinc tolerance in *Anthoxanthum odoratum*. *Heredity*, **32**, 287–97.

Geer, B.W., and Heinstra, P.W.H. (1990) Alcohol dehydrogenase and alcohol tolerance in *Drosophila melanogaster*. In *Evolutionary and Ecological Genetics of Drosophila* (eds. J.S.F. Barker, W.T. Starmer, and R.J. McIntyre). Plenum, New York.

Georghiou, G.P. (1972). The evolution of resistance to pesticides. *Annual Review of Ecology and Systematics*, **3**, 133–68.

Gibbs, H.L., and Grant, P.R. (1987). Oscillating selection on Darwin's finches. *Nature*, **327**, 511–3.

Gibson, J.B., and Wilks, A.V. (1988). The alcohol dehydrogenase polymorphism of *Drosophila melanogaster* in relation to environmental ethanol, ethanol tolerance and alcohol dehydrogenase activity. *Heredity*, **60**, 403–14.

Gibson, J.B., Lewis, N., Adena, M.A., and Wilson, S.R. (1979). Selection for ethanol tolerance in two populations of *Drosophila melanogaster* segregating for ADH allozymes. *Australian Journal of Biological Sciences*, **32**, 387–98.

Gibson, J.B., May, T.W., and Wilks, A.W. (1981). Genetic variation at the alcohol dehydrogenase locus in *Drosophila melanogaster* in relation to environmental variation: ethanol levels in breeding sites and allozyme frequencies. *Oecologia*, **51**, 191–8.

Giesel, J.T. (1986). Genetic correlation structure of life history variables in outbred, wild *Drosophila melanogaster*: effects of photoperiod regime. *American Naturalist*, **128**, 593–603.

Giesel, J.T., and Zettler, E.E. (1980). Genetic correlations of life historical parameters and certain fitness indices in *Drosophila melanogaster*. r_M, r_S, diet breadth. *Oecologia*, **47**, 299–302.

Giesel, J.T., Murphy, P.A., and Manlove, M.N. (1982). The influence of temperature on genetic interrelationships of life history traits in a population of *Drosophila melanogaster*. What tangled data sets we weave. *American Naturalist*, **119**, 464–79.

Gillespie, J.H. (1973). Polymorphism in a random environment. *Theoretical Population Biology*, **4**, 193–5.

Gillespie, J.H. (1976). A general model to account for enzyme variation in natural populations. II. Characterization of the fitness functions. *American Naturalist*, **110**, 809–21.

Gillespie, J.H. (1977). Natural selection for variances in offspring numbers: a new evolutionary principle. *American Naturalist*, **111**, 1010–4.

Gillespie, J.H. (1978). A general model to account for enzyme variation in natural populations. V. The SAS-CFF model. *Theoretical Population Biology*, **14**, 1–14.

Gillespie, J.H. (1981). Mutation modification in a random environment. *Evolution*, **35**, 468–76.

Gillespie, J.H., and Turelli, M. (1989). Genotype-environment interactions and the maintenance of polygenic variation. *Genetics*, **121**, 129–38.

Gionfriddo, M., Vigue, C., and Weisgram, P. (1979). Seasonal variation in the frequencies of the alcohol dehydrogenase isoalleles of *Drosophila*: correlation with environmental factors. *Genetical Research*, **34**, 317–9.

Glynn, P.W. (1988). El Nino—Southern oscillation 1982–1983. Nearshore population, community, and ecosystem responses. *Annual Review of Ecology and Systematics*, **19**, 309–45.

Goolish, E.M., and Burton, R.S. (1989). Energetics of osmoregulation in an intertidal copepod: effects of anoxia and lipid reserves on the pattern of free amino acid accumulation. *Functional Ecology*, **3**, 81–9.

Gordon, G., Brown, A.S., and Pulsford, T. (1988). A koala (*Phascolarctos cinereus* Goldfuss) population crash during drought and heatwave conditions in south-western Queensland. *Australian Journal of Ecology*, **13**, 451–61.

Grant, B.R., and Grant, P.R. (1989). Natural selection in a population of Darwin's finches. *American Naturalist*, **133**, 377–93.

Graubard, M.A. (1932). Inversion in *Drosophila melanogaster*. *Genetics*, **17**, 81–105.

Graves, J.E., and Somero, G.N. (1982). Electrophoretic and functional enzymic evolution in four species of eastern Pacific barracudas from different thermal environments. *Evolution*, **36**, 97–106.

Greenslade, P.J.M. (1983). Adversity selection and the habitat templet. *American Naturalist*, **122**, 352–65.

Greenway, H., and Munns, R. (1980). Mechanisms of salt tolerance in non-halophytes. *Annual Review of Plant Physiology*, **31**, 149–90.

Grell, R.F. (1978). A comparison of heat and interchromosomal effects on recombination and interference in *Drosophila melanogaster*. *Genetics*, **89**, 65–77.

Gressel, J. (1984). Evolution of herbicide-resistant weeds. In *Origins and development of adaptation*. (Ciba Foundation Symposium 102) pp. 73–93. Pitman Books, London.

Grime, J.P. (1965). Shade tolerance in flowering plants. *Nature*, **208**, 161–3.

Grime, J.P. (1977). Evidence for the existence of three primary strategies in plants and its relevance to ecological and evolutionary theory. *American Naturalist*, **111**, 1169–94.

Grime, J.P. (1979). *Plant strategies and vegetation processes*. John Wiley, Chichester.

Grime, J.P., Crick, J.L., and Rincon, J.E. (1986). The ecological significance of plasticity. In *Plasticity in plants* (eds. D.H. Jennings and A.J. Trewavas), pp. 5–19. Company of Biologists, Cambridge.

Groeters, F.R., and Dingle, H. (1988). Genetic and maternal influences on life history plasticity in milkweed bugs (*Oncopeltus*): response to temperature. *Journal of Evolutionary Biology*, **1**, 317–33.

Grosch, D.S., and Hopwood, L.E. (1979). *Biological effects of radiation*, 2nd edn. Academic Press, New York.

Gross, L.J. (1984). On the phenotypic plasticity of leaf photosynthetic capacity. In *Mathematical ecology* (eds. S.A. Levin and T.G. Hallam), pp. 2–13. Springer-Verlag, New York.

Grulke, N.E., and Bliss, L.C. (1988). Comparative life history characteristics of two high arctic grasses, Northwest Territories. *Ecology*, **69**, 484–96.

Hadley, N.F. (1977). Epicuticular lipids of the desert tenebrionid beetle, *Eleodes armata*: seasonal and acclimatory effects on composition. *Insect Biochemistry*, **7**, 277–83.

Haldane, J.B.S. (1932). *The causes of evolution*. Longmans, Green and Co., London.

Hale, H.B. (1970). Cross-adaptation. In *Physiology, environment and man* (eds. D.H.K. Lee and D. Minard), pp. 158–76. Academic Press, New York.

Haller, J., and Wittenberger, C. (1988). Biochemical energetics of hierarchy formation in *Betta splendens*. *Physiology and Behaviour* **43**, 447–50.

Hamilton, W.J., and Heppner, F. (1967). Radiant solar energy and the function of black homeotherm pigmentation: an hypothesis. *Science*, **155**, 196–7.

Hamrick, J.L. (1979). Genetic variation and longevity. In *Topics in plant population biology* (eds. O.T. Solbrig, S. Jain, G.B. Johnson and P.H. Raven), pp. 84–114. Columbia University Press, New York.

Hamrick, J.L., and Allard, R.W. (1972). Microgeographical variation in allozyme frequencies in *Avena barbata*. *Proceedings of the National Academy of Sciences, USA*, **69**, 2100–4.

Hanson, A.D., and Nelsen, C.E. (1980). Water: Adaptation of crops to drought-prone environments. In *The biology of crop productivity* (ed. P.S. Carlson), pp. 77–152. Academic Press, New York.

Hardin, R.T., and Bell, A.E. (1967). Two-way selection for body weight in *Tribolium* on two levels of nutrition. *Genetical Research*, **9**, 309–30.

Harper, J.L. (1977). *Population biology of plants*. Academic Press, London.

Harrison, R.G. (1980). Dispersal polymorphism in insects. *Annual Review of Ecology and Systematics*, **11**, 95–118.

Hartl, D.L., Dykhuizen, D.E., and Dean, A.M. (1985). Limits of adaptation: the evolution of selective neutrality. *Genetics*, **111**, 655–74.

Hawkins, A.J.S., Bayne, B.L., and Day, A.J. (1986). Protein turnover, physiological energetics and heterozygosity in the blue mussel, *Mytilus edulis*: the basis of variable age-specific growth. *Proceedings of the Royal Society of London, B*, **229**, 161–76.

Hawksworth, D.L., and Rose, F. (1976). *Lichens as pollution monitors*, Edward Arnold, London.

Heath, D.J. (1975). Colour, sunlight and internal temperatures in the land snail *Cepaea nemoralis* (L.). *Oecologia*, **19**, 29–38.

Hedrick, P.W. (1986). Genetic polymorphism in heterogeneous environments: a decade later. *Annual Review of Ecology and Systematics*, **17**, 535–66.

Hedrick, P.W., Ginevan, M.E., and Ewing, E.P. (1976). Genetic polymorphism in heterogeneous environments. *Annual Review of Ecology and Systematics*, **7**, 1–32.

Heslop-Harrison, J. (1964). Forty years of genecology. *Advances in Ecological Research*, **2**, 159–247.

Heuts, M.J. (1948). Adaptive properties of carriers of certain gene arrangements in *Drosophila pseudoobscura*. *Heredity*, **2**, 63–75.

Hickey, D.A., and McNeilly, T. (1975). Competition between metal tolerant and normal plant populations: a field experiment on normal soil. *Evolution*, **29**, 458–64.

Hilbish, T.J., and Koehn, R.K. (1985). Dominance in physiological phenotypes and fitness at an enzyme locus. *Science*, **229**, 52–4.

Hill, W.G. (1982). Predictions of response to artificial selection for new mutations. *Genetical Research*, **40**, 255–78.

Hinds, D.S., and Macmillen, R.E. (1985). Scaling energy metabolism and evaporative water loss in heteromyid rodents. *Physiological Zoology*, **58**, 282–98.

Hinrichsen, D. (1986). Multiple pollutants and forest decline. *Ambio* **15**, 258–65.

Hochachka, P.W. (1988). Channels and pumps-determinants of metabolic cold adaptation strategies. *Comparative Biochemistry and Physiology*, **90B**, 515–9.

Hochachka, P.W., and Somero, G.N. (1984). *Biochemical adaptation*. Princeton University Press, Princeton.

Hoenigsberg, H.F., Palomino, J.J., Hayes, M.J., Zandstra, I.Z., and Rolas, G.G. (1977). Population genetics in the American tropics. X. Genetic load differences in *Drosophila willistoni* from Columbia. *Evolution*, **31**, 805–11.

Hoffmann, A.A., and Cohan, F.M. (1987). Genetic divergence under uniform selection. III. Selection for knockdown resistance to ethanol in *Drosophila pseudoobscura* populations and their replicate lines. *Heredity*, **58**, 425–33.

Hoffmann, A.A., and Parsons, P.A. (1988). The analysis of quantitative variation in natural populations with isofemale strains. *Genetique, Selection, Evolution*, **20**, 87–98.

Hoffmann, A.A., and Parsons, P.A. (1989a). An integrated approach to environmental stress tolerance and life-history variation. Desiccation tolerance in *Drosophila*. *Biological Journal of the Linnean Society*, **37**, 117–36.

Hoffmann, A.A., and Parsons, P.A. (1989b). Selection for increased desiccation resistance in *Drosophila melanogaster*: additive genetic control and correlated responses for other stresses. *Genetics*, **122**, 837–45.

Hoffmann, A.A., and O'Donnell, S. (1990). Heritable variation in resource use in *Drosophila* in the field. In *Evolutionary and Ecological Genetics of Drosophila* (eds. J.S.F. Barker, W.T. Starmer, and R.J. McIntyre). Plenum, New York.

Hoffmann, A.A., Nielsen, K.M., and Parsons, P.A. (1984). Spatial variation of biochemical and ecological phenotypes in *Drosophila*—electrophoretic and quantitative variation. *Developmental Genetics*, **4**, 439–50.

Hogstad, O. (1987). It is expensive to be dominant. *Auk*, **104**, 333–6.

Holt, J.S., and Radosevich, S.R. (1983). Differential growth of two common groundsel (*Senecio vulgaris*) biotypes. *Weed Science*, **31**, 112–20.

Huether, C.A. (1969). Constancy of the pentamerous corolla phenotype in natural populations of *Linanthus*. *Evolution*, **23**, 572-88.

Huey, R.B., and Hertz, P.E. (1984). Is a jack-of-all-temperatures a master of none? *Evolution*, **38**, 441-4.

Huiskes, A.H.L., and Stienstra, A.W. (1985). Disasters and catastrophes in populations of *Halimione portulacoides*. In *Studies on plant demography: a festschrift for John L. Harper* (ed. J. White), pp. 83-93. Academic Press, London.

Hunter, A.S. (1964). Effects of temperature on *Drosophila*. I. Respiration of *Drosophila melanogaster* grown at different temperatures. *Comparative Biochemistry and Physiology*, **11**, 411-7.

Hunter, A.S. (1966). Effects of temperature on *Drosophila*. III. Respiration of *D. willistoni* and *D. hydei* grown at different temperatures. *Comparative Biochemistry and Physiology*, **19**, 171-77.

Hunter, A.S. (1968). Effects of temperature on *Drosophila*. IV. Adaptation of *D. immigrans*. *Comparative Biochemistry and Physiology*, **24**, 327-33.

Hutchinson, T.C. (1984). Adaptation of plants to atmospheric pollutants. In *Origins and development of adaptation* (CIBA Foundation Symposium 102), pp. 52-72. Pitman Books, London.

Istock, C.A., Zisfein, J., and Vavra, K.J. (1976). Ecology and evolution of the pitcher-plant mosquito. II. The substructure of fitness. *Evolution*, **30**, 535-47.

Ivanovici, A.M., and Wiebe, R.J. (1981). Towards a working 'definition' of 'stress': a review and critique. In *Stress effects on natural ecosystems* (eds. G.W. Barrett and R. Rosenberg), pp. 13-27. John Wiley, New York.

Jaenike, J. (1989). Genetic population structure of *Drosophila tripunctata*: patterns of variation and covariation of traits affecting resource use. *Evolution*, **43**, 1467-82.

Jain, S.K. (1969). Comparative ecogenetics of two *Avena* species occurring in central California. *Evolutionary Biology*, **3**, 73-118.

Jain, S.K. (1979). Adaptive strategies: polymorphism, plasticity and homeostasis. In *Topics in plant population biology* (eds. O.T. Solbrig, S.K. Jain, G.B. Johnson and P.H. Raven), pp. 160-87. Columbia University Press, New York.

Jarvinen, A., and Vaisanen, R.A. (1984). Reproduction of pied flycatchers (*Fidecula hypoleuca*) in good and bad breeding seasons in a northern marginal area. *Auk*, **101**, 439-50.

Jefferies, R.L., Jensen, A., and Bazely, D. (1983). The biology of the annual *Salicornia europaea* agg. at the limits of its range in Hudson Bay. *Canadian Journal of Botany*, **61**, 762-73.

Jinks, J.L., Perkins, J.M., and Pooni, H.S. (1973). The incidence of epistasis in normal and extreme environments. *Heredity*, **31**, 263-9.

Johns, D.M., and Miller, D.C. (1982). The use of bioenergetics to investigate the mechanisms of pollutant toxicity in crustacean larvae. In *Physiological mechanisms of marine pollutant toxicity* (eds. W.B. Vernberg, A. Calabrese, F.P. Thurberg and F.J. Vernberg), pp. 261-88. Academic Press, New York.

Johnson, G.R., and Frey, K.J. (1967). Heritabilities of quantitative attributes of oats (*Avena* sp.) at varying levels of environmental stress. *Crop Science*, **7**, 43-6.

Jokiel, P.L., and Coles, S.L. (1974). Effects of heated effluent on hermatypic corals at Kahe Point, Oahu. *Pacific Science*, **28**, 1–18.

Jones, J.S. (1982). Genetic differences in individual behaviour associated with shell polymorphism in the snail *Cepaea nemoralis*. *Nature*, **298**, 749–50.

Jones, J.S., Leith, B.H., and Rawlings, P. (1977). Polymorphism in *Cepaea*: a problem with too many solutions? *Annual Review of Ecology and Systematics*, **8**, 109–43.

Jonsson, P., Major, F., and Horst, P. (1988). Relationship between stress resistance and activation of hydrocorticosterone in mice. *Journal of Animal Breeding and Genetics*, **105**, 129–34.

Joshi, A., and Mueller, L.D. (1988). Evolution of higher feeding rate in *Drosophila* due to density-dependent natural selection. *Evolution*, **42**, 1090–3.

Juliano, S.A. (1986). Resistance to desiccation and starvation of two species of *Brachinus* (Coleoptera: Carabidae) from southeastern Arizona. *Canadian Journal of Zoology*, **64**, 73–80.

Kacser, H., and Burns, J.A. (1973). The control of flux. *Symposia of the Society for Experimental Biology*, **27**, 65–104.

Kacser, H., and Burns, J.A. (1981). The molecular basis of dominance. *Genetics*, **97**, 639–66.

Kahler, A.L., Allard, R.W., Krzakowa, M., Wehrhahn, C.F., and Nevo, E. (1980). Associations between isozyme phenotypes and environment in the slender wild oat (*Avena barbata*) in Israel. *Theoretical and Applied Genetics*, **56**, 31–47.

Kaitala, A. (1987). Dynamic life-history strategy of the waterstrider *Gerris thoracicus* as an adaptation to food and habitat variation. *Oikos*, **48**, 125–31.

Kaitala, A. (1988). Wing muscle dimorphism: two reproductive pathways of the waterstrider *Gerris thoracicus* in relation to habitat instability. *Oikos*, **53**, 222–8.

Kamping, A., and van Delden, W. (1978). The alcohol dehydrogenase polymorphism in populations of *Drosophila melanogaster*. II. The relation between ADH activity and adult mortality. *Biochemical Genetics*, **16**, 541–51.

Kaplan, R.H., and Cooper, W.J. (1984). The evolution of developmental plasticity in reproductive characteristics: an application of the 'adaptive coin-flipping' principle. *American Naturalist*, **123**, 393–410.

Kapoor, M., and Lewis, J. (1987). Heat shock induces peroxidase activity in *Neurospora crassa* and confers tolerance to oxidative stress. *Biochemical and Biophysical Research Communications*, **147**, 904–10.

Kearsey, M.J., and Barnes, B.W. (1970). Variation for metrical characters in *Drosophila* populations. II. Natural selection. *Heredity*, **25**, 11–21.

Keim, D.L., and Kronstad, W.E. (1979). Drought resistance and dryland adaptation in winter wheat. *Crop Science*, **19**, 574–6.

Ketzner, P.A., and Bradley, B.P. (1982). Rate of environmental change and adaptation in the copepod *Eurytemora affinis*. *Evolution*, **36**, 298–306.

Kimura, M. (1967). On the evolutionary adjustment of spontaneous mutation rates. *Genetical Research*, **9**, 23–34.

King, J.C., and Somme, L. (1958). Chromosomal analysis of the genetic factors for resistance to DDT in two resistant lines of *Drosophila melanogaster*. *Genetics*, **43**, 577–93.

Kingsolver, J.G., and Watt, W.B. (1983). Thermoregulatory strategies in *Colias* butterflies: thermal stress and the limits to adaptation in thermally varying environments. *American Naturalist*, **121**, 32–55.

Kinne, O. (1971). *Marine ecology Vol. I, Environmental factors*. Wiley, London.

Klerks, P.L., and Levinton, J.S. (1989). Rapid evolution of metal resistance in a benthic oligochaete inhabiting a metal-polluted site. *Biological Bulletin*, **176**, 135–41.

Knight, R. (1973). The relation between hybrid vigour and genotype-environment interactions. *Theoretical and Applied Genetics*, **43**, 311–8.

Koehn, R.K. (1978). Physiology and biochemistry of enzyme variation: the interface of ecology and population genetics. In *Ecological genetics: the interface* (ed. P.F. Brussard), pp. 51–72. Springer-Verlag, New York.

Koehn, R.K. and Bayne, B.L. (1989). Towards a physiological and genetical understanding of the energetics of the stress response. *Biological Journal of the Linnean Society*, **37**, 157–71.

Koehn, R.K., and Shumway, S.R. (1982). A genetic/physiological explanation for differential growth rate among individuals of the American oyster *Crassostrea virginica* (Gmelin). *Marine Biology Letters*, **3**, 35–42.

Koehn, R.K., Diehl, W.J., and Scott, T.M. (1988). The differential contribution by individual enzymes of glycolysis and protein catabolism to the relationship between heterozygosity and growth rate in the coot clam *Mulinia lateralis*. *Genetics*, **118**, 121–30.

Kohane, M.J., and Parsons, P.A. (1986). Environment-dependent fitness differences in *Drosophila melanogaster*: Temperature, domestication and the alcohol dehydrogenase locus. *Heredity*, **17**, 289–304.

Kohane, M.J., and Parsons, P.A. (1988). Domestication: evolutionary change under stress. *Evolutionary Biology*, **23**, 31–48.

Kohlmann, B., Nix, H., and Shaw, D.D. (1988). Environmental predictions and distributional limits of chromosomal taxa in the Australian grasshopper *Caledia captiva* (F.). *Oecologia*, **75**, 483–93.

Korkman, N. (1961). Selection for size in mice in different nutritional environments. *Heredity*, **47**, 342–56.

Lack, D. (1966). *Population studies of birds*. Oxford University Press, Oxford.

Lande, R. (1975). The maintenance of genetic variability by mutation in a polygenic character with linked loci. *Genetical Research*, **26**, 221–34.

Lande, R. (1976). Natural selection and random genetic drift in phenotypic evolution. *Evolution*, **30**, 314–34.

Lande, R. (1981). The minimum number of genes contributing to quantitative variation between and within populations. *Genetics*, **99**, 541–53.

Lande, R. (1983). The response to selection on major and minor mutations affecting a metrical trait. *Heredity*, **50**, 47–65.

Lande, R. (1988). Genetics and demography in biological conservation. *Science*, **241**, 1455–60.

Lande, R., and Barrowclough, G.F. (1987). Effective population size, genetic variation, and their use in population management. In *Viable populations for conservation* (ed. M.E. Soulé), pp. 87–173. Cambridge University Press, Cambridge.

Langer, I., Frey, K.J., and Bailey, T. (1979). Associations among productivity, production response, and stability indexes in oat varieties. *Euphytica*, **28**, 17–24.

Langridge, J. (1962). A genetic and molecular basis for heterosis in *Arabidopsis* and *Drosophila*. *American Naturalist*, **96**, 5–27.

Langridge, J. (1968). Thermal responses of mutant enzymes and temperature limits to growth. *Molecular and General Genetics*, **103**, 116–26.

Langridge, J., and Griffing, B. (1959). A study of high temperature lesions in *Arabidopsis thaliana*. *Australian Journal of Biological Sciences*, **12**, 117–35.

Laurie-Ahlberg, C.C., *et al.* (1985). Genetic variability of flight metabolism in *Drosophila melanogaster*. II. Relationship between power output and enzyme activity levels. *Genetics*, **111**, 845–68.

Leamy, L. (1984). Morphometric studies in inbred and hybrid house mice. V. Directional and fluctuating asymmetry. *American Naturalist*, **123**, 579–93.

Leamy, L., and Atchley, W. (1985). Directional selection and developmental stability: Evidence from fluctuating asymmetry of morphometric characters in rats. *Growth*, **49**, 8–18.

Leary, E.F., and Allendorf, F.W. (1989). Fluctuating asymmetry as an indicator of stress: Implications for conservation biology. *Trends in Ecology and Evolution*, **4**, 214–5.

Lee, B.T.O., and Parsons, P.A. (1968). Selection, prediction and response. *Biological Reviews*, **43**, 139–74.

Lerner, I.M. (1954). *Genetic homeostasis*. Oliver and Boyd, Edinburgh.

Le Rudelier, D., Strom, A.R., Dandekar, A.M., Smith, L.T., and Valentine, R.C. (1984). Molecular biology of osmoregulation. *Science*, **227**, 1065–8.

Levine, R.P. (1952). Adaptive responses of some third chromosome types of *Drosophila pseudoobscura*. *Evolution*, **6**, 216–33.

Levins, R. (1968). *Evolution in changing environments*. Princeton University Press, Princeton.

Levitt, J. (1980). *Responses of plants to environmental stress*. Academic Press, New York.

Lewontin, R.C. (1974). *The genetic basis of evolutionary change*. Columbia University Press, New York.

Lewontin, R.C. (1984). Detecting population differences in quantitative characters as opposed to gene frequencies. *American Naturalist*, **123**, 115–24.

Li, G.C., and Hahn, G.M. (1978). Ethanol-induced tolerance to heat and adriamycin. *Nature*, **274**, 699–701.

Li, G.C., and Laszlo, A. (1985). Thermotolerance in mammalian cells: A possible role for heat shock proteins. In *Changes in eukaryotic gene expression in response to environmental stress* (eds. B.G. Atkinson and D.B. Walden), pp. 227–54. Academic Press, London.

Lighton, J.R.B., and Bartholomew, G.A. (1988). Standard energy metabolism of a desert harvester ant, *Pogonomyrmex rugosus*: Effects of temperature, body mass, group size, and humidity. *Proceedings of the National Academy of Sciences, USA*, **85**, 4765–9.

Lima, S.L. (1986). Predation risk and unpredictable feeding conditions: determinants of body mass in birds. *Ecology*, **67**, 377–85.

Lindgren, D. (1972). The temperature influence on the spontaneous mutation rate. *Hereditas*, **70**, 165–78.

Lindquist, S. (1986). The heat-shock response. *Annual Review of Biochemistry*, **55**, 1151–91.

Lindquist, S., and Craig, E.A. (1988). The heat-shock proteins. *Annual Review of Genetics*, **22**, 631–77.

Lints, F.A., and Hoste, C. (1974). The Lansing effect revisited. I. Lifespan. *Experimental Gerontology*, **9**, 51–69.

Lints, F.A., Stoll, J., Gruwez, G., and Lints, C.V. (1979). An attempt to select for increased longevity in *Drosophila melanogaster*. *Gerontology*, **25**, 192–204.

Liu, E.H., Sharitz, R.R., and Smith, M.H. (1978). Thermal sensitivities of malate dehydrogenase isozymes in Typha. *American Journal of Botany*, **65**, 214–20.

Lively, C.M. (1986). Canalization versus developmental conversion in a spatially variable environment. *American Naturalist*, **128**, 561–72.

Loach, K. (1967). Shade tolerance in tree seedlings. *New Phytologist*, **66**, 607–21.

Lofsvold, D. (1986). Quantitative genetics of morphological differentiation in *Peromyscus*. I. Tests of the homogeneity of genetic covariance structure among species and subspecies. *Evolution*, **40**, 559–73.

Lo Gullo, M.A., and Salleo, S. (1988). Different strategies of drought resistance in three Mediterranean sclerophyllous trees growing in the same environmental conditions. *New Phytologist*, **108**, 267–76.

Loomis, W.F., and Wheeler, S.A. (1982). Chromatin-associated heat shock proteins of *Dictyostelium*. *Developmental Biology*, **90**, 412–8.

Lovegrove, B.G. (1986). The metabolism of social subterranean rodents: adaptation to aridity. *Oecologia*, **69**, 551–5.

Lucchesi. J.C., and Suzuki, D.T. (1968). The interchromosomal control of recombination. *Annual Review of Genetics*, **2**, 53–86.

Luckinbill, L.S., and Clare, M.J. (1985). Selection for lifespan in *Drosophila melanogaster*. *Heredity*, **55**, 9–18.

Luckinbill, L.S., Arking, R., Clare, M.J., Cirocco, W.C., and Buck, S. (1984). Selection for delayed senescence in *Drosophila melanogaster*. *Evolution*, **38**, 996–1003.

Luckinbill, L.S., Graves, J.L., Tomkin, A., and Sowirka, O. (1988). A qualitative analysis of some life-history correlates of longevity in *Drosophila melanogaster*. *Evolutionary Ecology*, **2**, 85–94.

Luckinbill, L.S., Grudzien, T.A., Rhine, S., and Weisman, G. (1989). The genetic basis of adaptation to selection for longevity in *Drosophila melanogaster*. *Evolutionary Ecology*, **3**, 31–9.

Lynch, M., and Gabriel, W. (1987). Environmental tolerance. *American Naturalist*, **129**, 283–303.

MacArthur, R.H., and Wilson, E.O. (1967). *The theory of island biogeography*. Princeton University Press, Princeton.

Macisaac, H.J., Hebert, P.D.N., and Schwartz, S.S. (1985). Inter- and intra-specific variation in acute thermal tolerance of *Daphnia*. *Physiological Zoology*, **58**, 350–5.

MacNair, M.R. (1981). Tolerance of higher plants to toxic materials. In *Genetic consequences of man-made change* (eds. J.A. Bishop and L.M. Cook), pp. 177–207. Academic Press, London.

MacNair, M.R. (1983). The genetic basis of copper tolerance in the yellow monkey flower *Mimulus guttatus*. *Heredity*, **50**, 283–93.

MacPhee, D.G. (1985). Indications that mutagenesis in *Salmonella* may be subject to catabolite repression. *Mutation Research*, **151**, 35–41.

Mahmoud, A., and Grime, J.P. (1974). A comparison of negative relative growth rates in shaded seedlings. *New Phytologist*, **73**, 1215–9.

Malik, R.C. (1984). Genetic and physiological aspects of growth, body composition and feed efficiency in mice: A review. *Journal of Animal Science*, **58**, 577–90.

Mansfield, T.A., and Freer-Smith, P.H. (1981). Effects of urban air pollution on plant growth. *Biological Reviews*, **56**, 343–68.

Marinkovic, D., and Ayala, F.J. (1986). Genetic variation for rate of development in natural populations of *Drosophila melanogaster*. *Genetika*, **71**, 123–32.

Marinkovic, D., Tucic, N., Moya, A., and Ayala, F.J. (1987). Genetic diversity and linkage disequilibrium in *Drosophila melanogaster* with different rates of development. *Genetics*, **117**, 513–20.

Marx, J.L. (1983). Surviving heat shock and other stresses. *Science*, **221**, 251–3.

Mather, K. (1943). Polygenic inheritance and natural selection. *Biological Reviews*, **18**, 32–64.

Mather, K. (1953). Genetic control of stability in development. *Heredity*, **7**, 297–336.

Mather, K. (1966). Variability and selection. *Proceedings of the Royal Society, Series B*, **164**, 328–40.

Mather, K. (1973). *Genetical structure of populations*. Chapman and Hall, London.

Mather, K., and Jinks, J.L. (1971). *Biometrical genetics: the study of continuous variation*. Chapman and Hall, London.

Matheson, A.C., and Parsons, P.A. (1973). The genetics of resistance to long-term exposure to CO_2 in *Drosophila melanogaster*, an environmental stress leading to anoxia. *Theoretical and Applied Genetics*, **42**, 261–68.

Mattson, W.J., and Haack, R.A. (1987). The role of drought in outbreaks of plant-eating insects. *Biological Reviews*, **37**, 110–8.

Maynard Smith, J. (1988). Selection for recombination in a polygenic model—the mechanism. *Genetical Research*, **51**, 59–63.

Mayr, E. (1963). *Animal species and evolution*. Belknap Press, Harvard.

Mayr, E. (1982). *The growth of biological thought: diversity, evolution and inheritance*. Belknap Press, Harvard.

McCarthy, J.C. (1980). Morphological and physiological effects of selection for growth rate in mice. In *Selection experiments in laboratory and domestic animals* (ed. A. Robertson), pp. 100–9. Commonwealth Agricultural Bureaux.

McClintock, B. (1951). Chromosome organization and genic expression. *Cold Spring Harbor Symposia in Quantitative Biology*, **16**, 13–47.

McClintock, B. (1984). The significance of responses of the genome to challenge. *Science*, **226**, 792–801.

McClure, P.A., and Randolph, J.C. (1980). Relative allocation of energy to growth and development of homeothermy in the eastern wood rat (*Neotoma floridana*) and hispid cotton rat (*Sigmodon hispidus*). *Ecological Monographs*, **50**, 199–219.

McGraw, J.B. (1987). Experimental ecology of *Dryas octopetala* ecotypes. IV. Fitness responses to reciprocal transplanting in ecotypes with differing plasticity. *Oecologia*, **73**, 465–8.

McKechnie, S.W., and Morgan, P. (1982). Alcohol dehydrogenase polymorphism of *Drosophila melanogaster*: aspects of alcohol and temperature variation in the larval environment. *Australian Journal of Biological Sciences*, **35**, 85–93.

McKechnie, S.W., and Geer, B.W. (1988). The epistasis of *Adh* and *Gpdh* allozymes in ethanol tolerance variation of *Drosophila melanogaster* larvae. *Genetical Research*, **52**, 179–84.

McKenney, C.L., and Dean, J.M. (1974). Effects of acute exposure to sublethal concentrations of cadmium on the thermal resistance of the mummichog. In *Thermal ecology* (eds. J. Gibbons and R. Sharitz), pp. 43–53. US Atomic Energy Commission.

McKenzie, J.A., and Clarke, G.M. (1988). Diazinon resistance, fluctuating asymmetry and fitness in the Australian sheep blowfly, *Lucilia cuprina*. *Genetics*, **120**, 213–20.

McKenzie, J.A., and Parsons, P.A. (1972). Alcohol tolerance: an ecological parameter in the relative success of *Drosophila melanogaster* and *Drosophila simulans*. *Oecologia*, **10**, 373–88.

McKenzie, J.A., and Parsons, P.A. (1974*a*). Microdifferentiation in a natural population of *Drosophila melanogaster* to alcohol in the environment. *Genetics*, **77**, 385–94.

McKenzie, J.A., and Parsons, P.A. (1974*b*). The genetic architecture of resistance to desiccation in populations of *Drosophila melanogaster* and *D. simulans*. *Australian Journal of Biological Science*, **27**, 441–56.

McKersie, B.D., and Hunt, L.A. (1987). Genotypic differences in tolerance of ice encasement, low temperature flooding, and freezing in winter wheat. *Crop Science*, **27**, 860–3.

McNab, B.K. (1973). Energetics and the distribution of vampires. *Journal of Mammology*, **54**, 131–44.

McNab, B.K. (1979). Climatic adaptation in the energetics of heteromyid rodents. *Comparative Biochemistry and Physiology*, **62A**, 813–20.

McNab, B.K. (1980). Food habits, energetics, and the population biology of mammals. *American Naturalist*, **116**, 106–24.

McNab, B.K. (1986). The influence of food habits on the energetics of eutherian mammals. *Ecological Monographs*, **56**, 1–19.

McNab, B.K., and Morrison, P. (1963). Body temperature and metabolism in subspecies of *Peromyscus* from arid and mesic environments. *Ecological Monographs*, **33**, 63–82.

McNaughton, S.J. (1974). Natural selection at the enzyme level. *American Naturalist*, **108**, 616–24.

Mearns, L.O., Katz, R.W., and Schneider, S.H. (1984). Extreme high-temperature events: changes in their probabilities with changes in mean temperature. *Journal of Climatology and Applied Meteorology*, **23**, 1601–13.

Mehlhorn, H., and Wellburn, A.R. (1987). Stress ethylene formation determines plant sensitivity to ozone. *Nature*, **327**, 417–8.

Menges, E.S., and Waller, D.M. (1983). Plant strategies in relation to elevation and light in floodplain herbs. *American Naturalist,* **122**, 454–73.

Michel, G.P.F., and Starka, J. (1987). Preferential synthesis of stress proteins in stationary *Zymomonas mobilis* cells. *FEMS Microbiology Letters,* **43**, 361–5.

Milkman, R.D. (1960). The genetic basis of natural variation. II. Analysis of a polygenic system in *Drosophila melanogaster. Genetics,* **45**, 377–91.

Milkman, R.D. (1965). The genetic basis of natural variation. VII. The individuality of polygenic combinations in *Drosophila. Genetics,* **52**, 789–99.

Milkman, R.D. (1979). The posterior crossvein in *Drosophila* as a model phenotype. In *Quantitative genetic variation* (eds. J.N. Thompson Jr. and J.M. Thoday), pp. 157–76. Academic Press, New York.

Mitton, J.B., and Grant, M.C. (1984). Associations among protein heterozygosity, growth rate, and developmental homeostasis. *Annual Review of Ecology and Systematics,* **15**, 479–99.

Mitton, J.B., Carey, C., and Kocher, T.D. (1986). The relation of enzyme heterozygosity to standard and active oxygen consumption and body size of tiger salamanders, *Ambystoma tigrinum. Physiological Zoology,* **59**, 574–82.

Mooney, H.A., and Billings, W.D. (1961). Comparative physiological ecology of arctic and alpine populations of *Oxyria digyna. Ecological Monographs,* **31**, 1–29.

Mooney, H.A., and Gulmon, S.L. (1979). Environmental and evolutionary constraints on the photosynthetic characteristics of higher plants. In *Topics in plant population biology* (ed. O.T. Solbrig), pp. 316–37. Macmillan, London.

Mooney, H.A., Vitousek, P.M., and Matson, P.A. (1987). Exchange of materials between terrestrial ecosystems and atmosphere. *Science,* **238**, 926–32.

Moreno, J., Carlson, A., and Alatalo, R.V. (1988). Winter energetics of coniferous forest tits *Paridae* in the north: the implications of body size. *Functional Ecology,* **2**, 163–70.

Morrison, W.W. and Milkman, R. (1978). Modification of heat resistance in *Drosophila* by selection. *Nature (London),* **273**, 49–50.

Mousseau, T.A., and Roff, D.A. (1987). Natural selection and the heritability of fitness components. *Heredity,* **59**, 181–97.

Moyed, H.S., Nguyen, T.T., and Bertrand, K.P. (1983). Multicopy Tn *10 tet* plasmids confer sensitivity to induction of *tet* gene expression. *Journal of Bacteriology,* **155**, 549–56.

Mueller, L.D. (1985). The evolutionary ecology of *Drosophila. Evolutionary Biology,* **19**, 37–98.

Mueller, L.D. (1988). Evolution of competitive ability in *Drosophila* by density-dependent natural selection. *Proceedings of the National Academy of Sciences, USA,* **85**, 4383–6.

Muller-Starck, G. (1985). Genetic differences between 'tolerant' and 'sensitive' beeches (*Fagus sylvatica* L.) in environmentally stressed adult forest stand. *Silvae Genetica,* **34**, 241–7.

Munday, A. (1961). Aspects of stress phenomena. *Symposia of the Society for Experimental Biology,* **15**, 168–89.

Murata, N., and Yamaya, J. (1984). Temperature-dependent phase behaviour of phosphatidylglycerols from chilling-sensitive and chilling-resistant plants.

Plant Physiology, **74**, 1016–24.

Murphy, P.A., Giesel, J.T., and Manlove, M.N. (1983). Temperature effects on life-history variation in *Drosophila simulans*. *Evolution*, **37**, 1181–92.

Myers, J.H., and Sabath, M.D. (1980). Genetics and phenotypic variability, genetic variance, and the success of establishment of insect introductions for the biological control of weeds. *Proceedings of the V International Symposium on the biological control of weeds* (ed. E.S. Delfosse), pp. 91–102. CSIRO, Melbourne.

Nagy, K.A. (1982). Ecological Energetics. In *Lizard ecology* (eds. R.B. Huey, E.R. Pianka and T.W. Schoener), pp. 24–54. Harvard University Press, Cambridge (Mass.).

Neel, J.V. (1941). A relation between larval nutrition and the frequency of crossing over in the third chromosome of *Drosophila melanogaster*. *Genetics*, **26**, 506–16.

Nei, M., Maruyama, T., and Chakraborty, R. (1975). The bottleneck effect and genetic variability in populations. *Evolution*, **29**, 1–10.

Nevo, E. (1978). Genetic variation in natural populations: patterns and theory. *Theoretical Population Biology*, **13**, 121–77.

Nevo, E. (1988). Genetic diversity in nature. Patterns and Theory. *Evolutionary Biology*, **23**, 217–46.

Nevo, E., and Shkolnik, A. (1974). Adaptive metabolic variation of chromosomal forms in male rats *Spalax*. *Experientia*, **30**, 724–6.

Nevo, E., Perl, T., Beiles, A., and Wool, D. (1981). Mercury selection of allozyme genotypes in shrimps. *Experientia*, **36**, 1152–4.

Nevo, E., Ben-Shlomo, R., and Lavie, B. (1984a). Mercury selection of allozymes in marine organisms: prediction and verification in nature. *Proceedings of the National Academy of Sciences, USA*, **81**, 1258–9.

Nevo, E., Beiles, A., and Ben-Shlomo, R. (1984b). The evolutionary significance of genetic diversity: ecological, demographic and life-history correlates. In *Evolutionary dynamics of genetic diversity* (ed. G.S. Mani), pp. 13–213. Springer Verlag, Berlin.

Newell, N.D. (1952). Periodicity in invertebrate evolution. *Journal of Palaentology*, **26**, 371–85.

Newell, R.C., and Branch, G.M. (1980). The influence of temperature on the maintenance of metabolic energy balance in marine invertebrates. *Advances in Marine Biology*, **17**, 329–96.

Newman, R.A. (1988a). Genetic variation for larval anuran (*Scaphiopus couchii*) development time in an uncertain environment. *Evolution*, **42**, 763–73.

Newman, R.A. (1988b). Adaptive plasticity in development of *Scaphiopus couchii* tadpoles in desert ponds. *Evolution*, **42**, 774–83.

Nicholls, M.K., and McNeilly, T. (1985). The performance of *Agrostis capillaris* L. genotypes, differing in copper tolerance, in ryegrass swards on normal soils. *New Phytologist*, **101**, 207–17.

Nielsen, K.M., Hoffmann, A.A., and McKechnie, S.W. (1985). Population genetics of the metabolically related Adh, Gpdh and Tpi polymorphisms in *Drosophila melanogaster*: II. Temporal and spatial variation in an orchard population. *Genetique, Selection, Evolution*, **17**, 41–58.

Nix, H.A. (1986). A biogeographic analysis of Australian elapid snakes. In *Atlas of elapid snakes of Australia* (ed. R. Longmore), pp. 4–15. Australian Fauna and Flora Series, no. 7, Bureau of Flora and Fauna, Canberra.

Nöthel, H. (1987). Adaptation of *Drosophila melanogaster* populations to high mutation pressure: evolutionary adjustment of mutation rates. *Proceedings of the National Academy of Sciences USA*, **84**, 1045–9.

Oakeshott, J.G., Gibson, J.B., Anderson, P.R., and Champ, A. (1980). Opposing modes of selection on the alcohol dehydrogenase locus in *Drosophila melanogaster*. *Australian Journal of Biological Sciences*, **33**, 105–14.

Oakeshott, J.G., Wilson, S.R., and Parnell, P. (1985). Selective effects of temperature on some enzyme polymorphisms in laboratory populations of *Drosophila melanogaster*. *Heredity*, **55**, 69–82.

Odum, E.P. (1983). *Basic ecology.* Holt-Saunders, Japan.

Odum, E.P. (1985). Trends expected in stressed ecosystems. *BioScience*, **35**, 419–22.

Odum, E.P., Finn, J.T., and Franz, E.H. (1979). Perturbation theory and the subsidy stress gradient. *BioScience*, **29**, 349–52.

Ogaki, M., and Nakashima-Tanaka, E. (1966). Inheritance of radioresistance in *Drosophila. Mutation Research*, **3**, 438–43.

Ogaki, M., Nakashima-Tanaka, E., and Murakami, S. (1967). Inheritance of ether resistance in *Drosophila melanogaster. Japanese Journal of Genetics*, **42**, 387–94.

Orians, G.H., and Solbrig, O.T. (1977). A cost–income model of leaves and roots with special reference to arid and semiarid areas. *American Naturalist*, **111**, 677–90.

Orzack, S.H. (1985). Population dynamics in variable environments. V. The genetics of homeostasis revisited. *American Naturalist*, **125**, 550–72.

Osgood, D.W. (1978). Effects of temperature on the development of meristic characters in *Natrix fasciata. Copeia* **1**, 33–47.

Osmond, C.B., *et al.* (1987). Stress physiology and the distribution of plants. *BioScience*, **37**, 38–48.

Palmer, A.R., and Strobeck, C. (1986). Fluctuating asymmetry: Measurement, analysis, pattern. *Annual Review of Ecology and Systematics*, **17**, 391–421.

Pani, S.N., and Lasley, J.F. (1972). *Genotype X environment interactions in animals.* Research Bulletin, Agricultural Experiment Station, University of Missouri—Columbia No. 992.

Pankakoski, E. (1985). Epigenetic asymmetry as an ecological indicator in muskrats. *Journal of Mammalogy*, **66**, 52–7.

Paquin, C.E., and Williamson, V.M. (1984). Temperature effects on the rate of Ty transposition. *Science*, **226**, 53–5.

Parker, M.A. (1984). Local food depletion and the foraging behavior of the specialized grasshopper, *Hesperotettix viridis. Ecology*, **65**, 824–35.

Paroda, R.S., and Hayes, J.D. (1971). An investigation of genotype–environment interactions for rate of ear emergence in spring barley. *Heredity*, **26**, 157–75.

Parry, M.L. (1978). *Climatic change, agriculture and settlement.* Dawson, Folkeston, England.

Parry, M.L., and Carter, T.R. (1985). The effect of climatic variations on agricultural risk. *Climatic Change*, **7**, 95–110.

Parsons, P.A. (1961). Fly size, emergence time and sternopleural chaeta number in *Drosophila*. *Heredity*, **16**, 455–73.

Parsons, P.A. (1962). Maternal age and developmental variability. *Journal of Experimental Biology*, **39**, 251–60.

Parsons, P.A. (1963). Migration as a factor in natural selection. *Genetica*, **33**, 184–206.

Parsons, P.A. (1964). Parental age and the offspring. *Quarterly Review of Biology*, **39**, 258–75.

Parsons, P.A. (1971). Extreme-environment heterosis and genetic loads. *Heredity*, **26**, 579–83.

Parsons, P.A. (1974). Genetics of resistance to environmental stresses in *Drosophila* populations. *Annual Review of Genetics*, **7**, 239–65.

Parsons, P.A. (1977). Genes, behavior, and evolutionary processes: the genus *Drosophila*. *Advances in Genetics*, **19**, 1–32.

Parsons, P.A. (1980). Adaptive strategies in natural populations of *Drosophila*: Ethanol tolerance, desiccation resistance, and development times in climatically optimal and extreme environments. *Theoretical and Applied Genetics*, **57**, 257–66.

Parsons, P.A. (1982*a*). Evolutionary ecology of Australian *Drosophila*: a species analysis. *Evolutionary Biology*, **14**, 297–350.

Parsons, P.A. (1982*b*). Acetic acid vapour as a resource and stress in *Drosophila*. *Australian Journal of Zoology*, **30**, 427–33.

Parsons, P.A. (1983). *The evolutionary biology of colonizing species*. Cambridge University Press, New York.

Parsons, P.A. (1987*a*). Features of colonizing animals: Phenotypes and genotypes. In *Colonization, succession and stability* (eds. A.J. Gray, M.J. Crawley and P.J. Edwards), pp. 133–154. Blackwell Scientific Publications, Oxford.

Parsons, P.A. (1987*b*). Evolutionary rates under environmental stress. *Evolutionary Biology*, **21**, 311–47.

Parsons, P.A. (1988). Evolutionary rates: effects of stress upon recombination. *Biological Journal of Linnean Society*, **35**, 49–68.

Parsons, P.A. (1989*a*). Acetaldehyde utilization in *Drosophila*: an example of hormesis. *Biological Journal of the Linnean Society*, **37**, 183–89.

Parsons, P.A. (1989*b*). Environmental stresses and conservation of natural populations. *Annual Review of Ecology and Systematics*, **20**, 29–49.

Parsons, P.A. (1990). Fluctuating asymmetry: an epigenetic measure of stress. *Biological Reviews*, **63**, 131–45.

Parsons, P.A., and Stanley, S.M. (1981). Domesticated and widespread species. In *The genetics and biology of Drosophila, Volume 3a* (eds. M. Ashburner, H.L. Carson and J.N. Thompson Jr), pp. 349–93. Academic Press, London.

Parsons, P.A., MacBean, I.T., and Lee, B.T.O. (1969). Polymorphism in natural populations for genes controlling radioresistance in *Drosophila*. *Genetics*, **61**, 211–18.

Partridge, G.G. (1979). Relative fitness of genotypes in a population of *Rattus norvegicus* polymorphic for warfarin resistance. *Heredity*, **43**, 239–46.

Partridge, L., and Harvey, P.H. (1988). The ecological context of life history evolution. *Science*, **241**, 1449–55.

Pederson, D.G. (1968). Environmental stress, heterozygote advantage and genotype–environment interaction in *Arabidopsis*. *Heredity*, **23**, 127–38.

Pelham, H.R.B. (1986). Speculations on the functions of the major heat shock and glucose-regulated proteins. *Cell,* **46**, 959–61.

Peng, T.X., Moya, A., and Ayala, F.J. (1986). Irradiation-resistance conferred by superoxide dismutase: possible adaptive role of a natural polymorphism in *Drosophila melanogaster. Proceedings of the National Academy of Sciences, USA,* **83**, 684–7.

Peters, R.L., and Darling, J.D.S. (1985). The greenhouse effect and nature reserves. *BioScience,* **35**, 707–17.

Peterson, C.H., and Black, R. (1988). Density-dependent mortality caused by physical stress interacting with biotic history. *American Naturalist,* **131**, 257–70.

Peterson, N.S., Moller, G., and Mitchell, H.K. (1979). Genetic mapping of the coding regions for three heat-shock proteins in *Drosophila melanogaster. Genetics,* **92**, 891–902.

Pfreim, P., and Sperlich, D. (1982). Wild O-chromosomes of *Drosophila subobscura* from different geographic regions have different effects on viability. *Genetica,* **60**, 49–59.

Piazza, A., Menozzi, P., and Cavalli-Sforza, L.L. (1981). Synthetic gene frequency maps of man and selective effects of climate. *Proceedings of the National Academy of Sciences, USA,* **78**, 2638–42.

Plapp, F.W. (1976). Biochemical genetics of insecticide resistance. *Annual Review of Entomology,* **21**, 179–97.

Plough, H.H. (1917). The effect of temperature on crossing over in *Drosophila. Journal of Experimental Zoology,* **24**, 148–209.

Polis, G.A. (1988). Foraging and evolutionary responses of desert scorpions to harsh environmental periods of food stress. *Journal of Arid Environments,* **14**, 123–34.

Pough, F.H. (1989). Organismal performance and darwinian fitness: approaches and interpretations. *Physiological Zoology,* **62**, 199–236.

Price, T.D., Grant, P.R., Gibbs, H.L., and Boag, P.T. (1984). Recurrent patterns of natural selection in a population of Darwin's finches. *Nature,* **309**, 787–89.

Pulich, W.M. (1974). Resistance to high oxygen tension, streptonigrin, and ultraviolet irradiation in the green alga *Chlorella sorokiniana* strain ORS. *Journal of Cell Biology,* **62**, 904–7.

Quinn, P.J. (1989). Principles of membrane stability and phase behavior under extreme conditions. *Journal of Bioenergetics and Biomembranes,* **21**, 3–19.

Raison, J.K., Chapman, E.A., Wright, I.C., and Jacobs, S.W.L. (1979). Membrane lipid transitions: their correlation with the climatic distributions of plants. In *Low temperature stress in crop plants: the role of the membrane* (eds. J.M. Lyons, D. Graham, and J.K. Raison), pp. 177–86. Academic Press, New York.

Ramsay, N. (1988). A mutant in a major heat shock protein of *Escherichia coli* continues to show inducible thermotolerance. *Molecular and General Genetics* **211**, 332–34.

Raup, D.M. (1981). Introduction: What is a crisis? In *Biotic crises in ecological and evolutionary time* (ed. M.H. Nitecki), pp. 1–12. Academic Press, New York.

Raup, D.M. (1986). Biological extinction in earth history. *Science,* **231**, 1528–33.

Raup, D.M., and Boyajian, G.E. (1988). Patterns of generic extinction in the fossil record. *Paleobiology*, **14**, 109–25.

Raup, D.M., and Sepkoski, J.J. (1984). Periodicity of extinctions in the geologic past. *Proceedings of the National Academy of Sciences, USA*, **81**, 801–5.

Reeve, E.C.R. (1960). Some genetic tests on asymmetry of sternopleural chaeta number in *Drosophila*. *Genetical Research*, **1**, 151–72.

Rehfeldt, G.E. (1979). Ecotypic differentiation in populations of *Pinus monticola* in North Idaho—myth or reality? *American Naturalist*, **114**, 627–36.

Reznick, D. (1985). Costs of reproduction: an evaluation of the empirical evidence. *Oikos*, **44**, 257–67.

Rich, S.S., and Bell, A.E. (1980). Genotype-environment interaction effects in long-term selected populations of *Tribolium*. *Journal of Heredity*, **71**, 319–22.

Richards, R.A. (1978). Genetic analysis of drought stress response in rapeseed (*Brassica campestris* and *B. napus*). I. Assessment of environments for maximum selection response in grain yield. *Euphytica*, **27**, 609–15.

Richardson, A.M.M. (1974). Differential climatic selection in natural population of land snail *Cepaea nemoralis*. *Nature*, **247**, 572–3.

Riska, B., Prout, T., and Turelli, M. (1989). Laboratory estimates of heritabilities and genetic correlations in nature. *Genetics*, **123**, 865–71.

Robson, M.J., and Jewiss, O.R. (1968). A comparison of British and North African varieties of tall fesue (*Festuca arundinacea*). *Journal of Applied Ecology*, **5**, 179–90.

Root, T. (1988*a*). Environmental factors associated with avian distributional limits. *Journal of Biogeography*, **15**, 489–505.

Root, T. (1988*b*). Energy constraints on avian ranges. *Ecology*, **69**, 330–9.

Rose, M.R. (1982). Antagonistic pleiotropy, dominance, and genetic variation. *Heredity*, **48**, 63–78.

Rose, M.R. (1984). Genetic covariation in *Drosophila* life history: untangling the data. *American Naturalist*, **123**, 565–9.

Rose, M.R., and Charlesworth, B. (1981). Genetics of life history in *Drosophila melanogaster*. I. Sib analysis of adult females. *Genetics*, **97**, 173–86.

Roskaft, E., Jarvi, T., Bakken, M., Bech, C., and Reinertsen, R.E. (1986). The relationship between social status and metabolic rate in great tits (*Parus major*) and pied flycatchers (*Ficedula hypoleuca*). *Animal Behaviour*, **34**, 838–42.

Roush, R.T., and McKenzie, J.A. (1987). Ecological genetics of insecticide and acaricide resistance. *Annual Review of Entomology*, **32**, 361–80.

Roush, R.T., and Plapp, F.W. (1982). Effects of insecticide resistance on biotic potential of the house fly (Diptera: Muscidae). *Journal of Economic Entomology*, **75**, 708–13.

Ruban, P.S., Cunningham, E.P., and Sharp, P.M. (1988). Heterosis × nutrition interaction in *Drosophila melanogaster*. *Theoretical and Applied Genetics*, **76**, 136–42.

Rumbaugh, M.D., Asay, K.H., and Johnson, D.A. (1984). Influence of drought stress on genetic variances of alfalfa and wheatgrass seedlings. *Crop Science*, **24**, 297–303.

Ryder, E.J. (1973). Selecting and breeding plants for increased resistance to air pollutants. In *Air pollution damage and vegetation* (ed. J.A. Naegle), pp. 75–84. American Chemical Society, Washington, D.C.

Sacher, G.A., and Duffy, P.H. (1979). Genetic relation of life-span to metabolic rate for inbred mouse strains and their hybrids. *Federation Proceedings*, **38**, 184–8.

Salt, G. (1984). *Ecology and evolutionary biology*. University of Chicago Press, Chicago.

Sampsell, B., and Sims, S. (1982). Effect of *adh* genotype and heat stress on alcohol tolerance in *Drosophila melanogaster*. *Nature*, **296**, 853–5.

Sankaranarayanan, K. (1982). *Genetic effects of ionizing radiation in multicellular eukayotes and the assessment of genetic radiation hazards in man*. Elsevier Biomedical Press, Amesterdam.

Sarukhan, J., Martinez-Ramos, M., and Pinero, D. (1984). The analysis of demographic variability at the individual level and its population consequences. In *Perspectives on plant population ecology* (eds. R. Dirzo and J. Sarukhan), pp. 83–106. Sinauer, Sunderland.

Sauer, J.R. (1985). Mortality associated with severe weather in a northern population of cotton rats. *American Midland Naturalist*, **113**, 188–9.

Scheiner, S.M., and Goodnight, C.J. (1984). The comparison of phenotypic plasticity and genetic variation in populations of the grass *Danthonia spicata*. *Evolution*, **38**, 845–55.

Scheiner, S.M., Caplan, R.L., and Lyman, R.F. (1989). A search for trade-offs among life history traits in *Drosophila melanogaster*. *Evolutionary Ecology*, **3**, 51–63.

Schimke, R.T., Sherwood, S.W., Hill, A.B., and Johnston, R.N. (1986). Over-replication and recombination of DNA in higher eukaryotes: potential consequences and biological implications. *Proceedings of the National Academy of Sciences, USA*, **83**, 2157–61.

Schlesinger, M.J., Ashburner, M., and Tissieres, A. (eds.) (1982). *Heat-shock: From bacteria to man*. Cold Spring Harbor Laboratory Press, Cold Spring Habor, New York.

Schlichting, C.D. (1986). The evolution of phenotypic plasticity in plants. *Annual Review of Ecology and Systematics*, **17**, 667–93.

Schmalhausen, I.I. (1949). *Factors of evolution*. Blakiston, Philadelphia.

Schmidt-Nielsen, K. (1984). *Scaling: Why is animal size so important?* Cambridge University Press, Cambridge.

Schnee, F.B., and Thompson, J.N. (1984) Conditional polygenic effects in the sternopleural bristle system of *Drosophila melanogaster*. *Genetics*, **108**, 409–24.

Schneider, S.H. (1989). The greenhouse effect: Science and policy. *Science*, **243**, 772–81.

Schneider, S.H., and Lander, R. (1984). *The co-evolution of climate and life*. Sierra Books, San Francisco.

Scholz, F., and Bergmann, F. (1984). Selection pressure by air pollution as studied by isozyme-gene-systems in Norway spruce exposed to sulphur dioxide. *Silvae Genetica*, **33**, 238–41.

Schonewald-Cox, C.M., Chambers, S.M., MacBryde, B., and Thomas, W.L. (eds.) (1983). *Genetics and conservation: A reference for managing wild animal and plant populations*. Benjamin Cummings, Menlo Park.

Schultz, J., and Redfield, H. (1951). Interchromosomal effects on crossing over in *Drosophila*. *Cold Spring Harbor Symposia on Quantitative Biology*, **16**, 175–97.

Sciulli, P.W., Doyle, W.J., Kelley, C., Siegel, P., and Siegel, M.I. (1979). The interaction of stressors in the induction of increased levels of fluctuating asymmetry in the laboratory rat. *American Journal of Physical Anthropology*, **50**, 279–84.

Sedcole, J.R. (1981). A review of the theories of heterosis. *Egyptian Journal of Genetics and Cytology*, **10**, 117–46.

Sedensky, M.M., and Meneely, P.M. (1987). Genetic analysis of halothane sensitivity in *Caenorhabditis elegans*. *Science*, **236**, 952–4.

Selye, H. (1950). *The physiology and pathology of exposure to stress*. Acta Inc., Montreal.

Selye, H. (1952). *The story of the adaptation syndrome*. Acta Inc., Montreal.

Selye, H. (1956). *The stress of life*. McGraw-Hill, New York.

Service, P.M. (1987). Physiological mechanisms of increased stress resistance in *Drosophila melanogaster* selected for postponed senescence. *Physiological Zoology*, **60**, 321–6.

Service, P.M. (1989). The effect of mating status on lifespan, egg laying, and starvation resistance in *Drosophila melanogaster* in relation to selection on longevity. *Journal of Insect Physiology*, **35**, 447–52.

Service, P.M., and Rose, M.R. (1985). Genetic covariation among life-history components: the effect of novel environments. *Evolution*, **39**, 943–5.

Service, P.M., Hutchinson, E.W., Mackinley, M.D., and Rose, M.R. (1985). Resistance to environmental stress in *Drosophila melanogaster* selected for postponed senescence. *Physiological Zoology*, **58**, 380–9.

Service, P.M., Hutchinson, E.W., and Rose, M.R. (1988). Multiple genetic mechanisms for the evolution of senescence in *Drosophila melanogaster*. *Evolution*, **42**, 708–16.

Shaw, J. (1988). Genetic variation for tolerance to copper and zinc within and among populations of the moss, *Funaria hygrometrica* Hedw. *New Phytologist*, **109**, 211–22.

Sibly, R.M., and Calow, P. (1986). *Physiological ecology of animals: an evolutionary approach*. Blackwell Scientific Publications, Oxford.

Sibly, R.M., and Calow, P. (1989). A life-cycle theory of responses to stress. *Biological Journal of the Linnean Society*, **37**, 101–16.

Siegel, M.I., Doyle, W.J., and Kelley, C. (1975). Heat stress, fluctuating asymmetry and prenatal selection in the laboratory rat. *American Journal of Physical Anthropology*, **46**, 121–6.

Sierra, L.M., Comendador, M.A., and Aguirrezabalaga, I. (1989). Mechanisms of resistance to acrolein in *Drosophila melanogaster*. *Genetique, Selection, Evolution*, **21**, 427–36.

Silander, J.A., and Antonovics, J. (1979). The genetic basis of the ecological amplitude of *Spartina patens*. I. Morphometric and physiological traits. *Evolution*, **33**, 1114–27.

Simberloff, D. (1988). The contribution of population and community biology to conservation science. *Annual Review of Ecology and Systematics*, **19**, 473–511.

Simon, J-P. (1979). Adaptation and acclimation of higher plants at the enzyme level: latitudinal variation of thermal properties of NAD malate dehydrogenase in *Lathyrus japonicus* Wild. (Leguminosae). *Oecologia*, **39**, 273–87.

Simon, J-P., Potuin, C., and Blanchard, M-H. (1983). Thermal adaptation and acclimation of higher plants at the enzyme level: kinetic properties of NAD malate dehydrogenase and glutamate oxaloacetate transaminase in two genotypes of *Arabidopsis thaliana* (Brassicaceae). *Oecologia*, **60**, 143–8.

Simon, J-P., Charest, C., and Peloquin, M-J. (1986). Adaptation and acclimation of higher plants at the enzyme level: kinetic properties of NAD malate dehydrogenase in three species of *Viola*. *Journal of Ecology*, **74**, 19–32.

Smith, M.H., Smith, M.W., Scott, S.L., Liu, E.H., and Jones, J.C. (1983). Rapid evolution in a post-thermal environment. *Copeia*, **1**, 193–7.

Smith-Gill, S.J. (1983). Developmental plasticity: developmental conversion versus phenotypic modulation. *American Zoologist*, **23**, 47–55.

Snaydon, R.W., and Bradshaw, A.D. (1961). Differential response to calcium within the species *Festuca ovina* L. *New Phytologist*, **60**, 219–34.

Sohal, R.S. (1986). The rate of living theory: a contemporary interpretation. In *Insect aging* (eds. K.-G. Collatz and R.S. Sohal), pp. 23–44. Springer-Verlag, Berlin.

Sohal, R.S., Farmer, K.J., and Allen, R.G. (1987). Correlates of longevity in two strains of the housefly, *Musca domestica*. *Mechanisms of Ageing and Development*, **40**, 171–9.

Somero, G.N. (1978). Temperature adaptation of enzymes: Biological optimization through structure–function compromises. *Annual Review of Ecology and Systematics*, **9**, 1–29.

Somero, G.N. (1986). Protein adaptation and biogeography: threshold effects on molecular evolution. *Trends in Ecology and Evolution*, **1**, 124–7.

Soulé, M.E. (1967). Phenetics of natural populations. II. Asymmetry and evolution in a lizard. *American Naturalist*, **101**, 141–60.

Soulé, M.E. (1973). The epistasis cycle: A theory of marginal populations. *Annual Review of Ecology and Systematics*, **4**, 165–87.

Soulé, M.E. (1987). *Viable populations for conservation*. Cambridge University Press, Cambridge.

Soulé, M.E., and Baker, B. (1968). Phenetics of natural populations. IV. The population asymmetry parameter in the butterfly, *Coenonympha tullia*. *Heredity*, **23**, 611–14.

Southwood, T.R.E. (1962). Migration of terrestrial arthropods in relation to habitat. *Biological Reviews*, **37**, 171–214.

Southwood, T.R.E. (1977). Habitat, the templet for ecological strategies. *Journal of Animal Ecology*, **46**, 337–65.

Southwood, T.R.E. (1988). Tactic, strategies and templets. *Oikos*, **52**, 3–18.

Spassky, B. (1951). Effect of temperature and moisture content of the nutrient medium on the viability of chromosomal types in *Drosophila pseudoobscura*. *American Naturalist*, **85**, 177–80.

Sperlich, D., Feuerbach-Mravlag, H., Lange, P., Michaelidis, A., and Pentzos-Daponte, A. (1977). Genetic load and viability distribution in central and marginal populations of *Drosophila subobscura*. *Genetics*, **86**, 835–48.

Spotila, J.R., Standora, E.A., Easton, D.P., and Rutledge, P.S. (1989). Bioenergetics, behavior, and resource partitioning in stressed habitats: biophysical and molecular approaches. *Physiological Zoology*, **62**, 253–85.

Stanley, S.M. (1984). Marine mass extinctions: a dominant role for temperature. *Extinctions* (ed. M.H. Nitecki), pp. 69–117. University of Chicago Press, Chicago.

Stanley, S.M., and Parsons, P.A. (1981). The response of the cosmopolitan species, *Drosophila melanogaster*, to ecological gradients. *Proceedings of the Ecological Society of Australia*, **11**, 121–30.

Stearns, S.C. (1976). Life-history tactics: A review of the ideas. *Quarterly Review of Biology*, **51**, 3–47.

Stearns, S.C. (1983). The genetic basis of differences in life-history traits among six populations of mosquitofish (*Gambusia affinis*) that shared ancestors in 1905. *Evolution*, **37**, 618–27.

Stephanou, G., and Alahiotis, S.N. (1986). Adaptive significance of the action of the *Drosophila melanogaster* alcohol dehydrogenase locus through the heat shock protein system. *Genetica*, **69**, 59–68.

Stephanou, G., Alahiotis, S.N., Christodoulou, C., and Marmaras, V.J. (1983). Adaptation of *Drosophila* to temperature: heat-shock proteins and survival in *Drosophila melanogaster*. *Developmental Genetics*, **3**, 299–308.

Strand, D.J., and McDonald, J.F. (1985). Copia is transcriptionally responsive to environmental stress. *Nucleic Acids Research*, **13**, 4401–10.

Strong, D.R. (1983). Natural variability and the manifold mechanisms of ecological communities. *American Naturalist*, **122**, 636–60.

Sultan, S.E. (1987). Evolutionary implications of phenotypic plasticity in plants. *Evolutionary Biology*, **21**, 127–78.

Symeonidis, L., McNeilly, T., and Bradshaw, A.D. (1985). Differential tolerance of three cultivars of *Agrostis capillaris* L. to cadmium, copper, lead, nickel and zinc. *New Phytologist*, **101**, 309–15.

Tabachnick, W.J., and Powell, J.R. (1977). Adaptive flexibility of 'marginal' versus 'central' populations of *Drosophila willistoni*. *Evolution*, **31**, 692–4.

Takahashi, F. (1977). Generation carryover of a fraction of population members as an animal adaptation to unstable environmental conditions. *Resources, Populations, Ecology*, **18**, 235–42.

Tantawy, A.O., and Mallah, G.S. (1961). Studies on natural populations of *Drosophila*. I. Heat resistance and geographical variation in *Drosophila melanogaster* and *D. simulans*. *Evolution*, **15**, 1–14.

Tauber, C.A., and Tauber, M.J. (1982). Evolution of seasonal adaptations and life history traits in *Chrysopa*: response to diverse selection pressures. In *Evolution and genetics of life-histories* (eds. H. Dingle and J.P. Hegmann), pp. 51–72. Springer, New York.

Tauber, M.J., Tauber, C.A., and Masaki, S. (1986). *Seasonal adaptations of insects*. Oxford University Press, New York.

Taylor, G.E. (1978). Genetic analysis of ecotypic differentiation within an annual plant species *Geranium carolinianum* L. in response to sulphur dioxide. *Botanical Gazette*, **139**, 362–8.

Taylor, G.E., and Murdy, W.H. (1975). Population differentiation of an annual plant species *Geranium carolinianum* L. in response to sulphur dioxide. *Botanical Gazette*, **136**, 212–5.

Templeton, A.R., and Johnston, J.S. (1982). Life history evolution under pleiotropy and K-selection in a natural population of *Drosophila melanogaster*. In *Ecological genetics and evolution: The cactus-yeast-Drosophila model system* (eds. J.S.F. Barker and W.T. Starmer), pp. 225–39. Academic Press, New York.

Templeton, A.R., and Johnston, J.S. (1988). The measured genotype approach to ecological genetics. In *Population genetics and evolution* (ed. G. de Jong), pp. 138–46. Springer-Verlag, Berlin.

Theede, H., Ponat, A., Hiroki, K., and Schlieper, C. (1969). Studies on the resistance of marine bottom invertebrates to oxygen-deficiency and hydrogen sulphide. *Marine Biology*, **2**, 325–37.

Thoday, J.M. (1956). Balance, heterozygosity and developmental stability. *Cold Spring Harbor Symposia in Quantitative Biology*, **21**, 318–26.

Thoday, J.M. (1958). Homeostasis in a selection experiment. *Heredity*, **12**, 401–15.

Thomson, J.A. (1971). Associations of karyotype with body weight and resistance to desiccation in *Drosophila pseudoobscura*. *Canadian Journal of Genetics and Cytology*, **13**, 63–9.

Tilling, S.M. (1983). An experimental investigation of the behaviour and mortality of artificial and natural morphs of *Cepaea nemoralis* (L.). *Biological Journal of the Linnean Society*, **19**, 35–50.

Timofeeff-Ressovsky, N.W. (1940). Mutations and geographical variation. In *The new systematics* (ed. J. Huxley), pp. 73–136. Clarendon Press, Oxford.

Ting, I.P., and Duggar, W.M. (1968). Non-autotrophic carbon dioxide metabolism in cacti. *Botanical Gazette*, **129**, 9–15.

Torres, J.J., and Somero, G.N. (1988). Metabolism, enzymic activities and cold adaptation in Antarctic mesopelagic fishes. *Marine Biology*, **98**, 169–80.

Tsuji, J. (1988). Seasonal profiles of standard metabolic rate of lizards (*Sceloporus occidentalis*) in relation to latitude. *Physiological Zoology*, **61**, 230–40.

Tucic, N. (1979). Genetic capacity for adaptation to cold resistance at different developmental stages of *Drosophila melanogaster*. *Evolution*, **33**, 350–8.

Turelli, M. (1984). Heritable genetic variation via mutation-selection balance: Lerch's zeta meets the abdominal bristle. *Theoretical Population Biology*, **25**, 138–93.

Turelli, M. (1986). Gaussian versus non-Gaussian genetic analyses of polygenic mutation-selection balance. In *Evolutionary processes and rheory* (eds. S. Karlin and E. Nevo), pp. 607–28. Academic Press, New York.

Turelli, M., and Barton, N. (1990). Dynamics of polygenic characters under selection. *Genetics* (in press).

Turesson, G. (1922). The genotypic response of the plant species to the habitat. *Hereditas*, **3**, 211–350.

Turin, H., and Den Boer, P.J. (1988). Changes in the distribution of carabid beetles in the Netherlands since 1880. II. Isolation of habitats and long-term trends in the occurrence of carabid species with different powers of dispersal (Coleoptera, Carabidae). *Biological Conservation*, **44**, 179–200.

Turner, H.G., and Schleger, A.V. (1960). The significance of coat type in cattle. *Australian Journal of Agricultural Research*, **11**, 645–663.

Turner, R.G. (1969). Heavy-metal tolerance in plants. In *Ecological aspects of mineral nutrition in plants* (ed. I.H. Rorison), pp. 399–410. Blackwell Scientific, Oxford.

Valentine, D.W., and Soulé, M.E. (1973). Effect of p,p′-DDT on developmental stability of pectoral fin rays in the grunion, *Leuresthes tenuis*. *Fishery Bulletin*, **71**, 921–6.

Valentine, D.W., Soulé, M.E., and Samollow, P. (1973). Asymmetry analysis in fishes: a possible statistical indicator of environmental stress. *Fishery Bulletin*, **71**, 357–70.

Van Noordwijk, A.J., Van Balen, J.H., and Scharloo, W. (1988). Heritability of body size in a natural population of the great tit (*Parus major*) and its relation to age and environmental conditions during growth. *Genetical Research*, **51**, 149–62.

Van Valen, L. (1962). A study of fluctuating asymmetry. *Evolution*, **16**, 125–42.

Van Valen, L., Levine, L., and Beardmore, J.A. (1962). Temperature sensitivity of chromosomal polymorphism in *Drosophila pseudoobscura*. *Genetica*, **33**, 113–27.

Van Vleck, L.D. (1963). Genotype and environment in sire evaluation. *Journal of Dairy Science*, **46**, 983–7.

Van Waarde, A. (1988). Operation of the purine nucleotide cycle in animal tissues. *Biological Reviews*, **63**, 259–98.

Venable, D.L. (1984). Using intraspecific variation to study the ecological significance and evolution of plant life-histories. In *Perspectives on plant population ecology* (eds. R. Dirzo and J. Sarukhan), pp. 166–87. Sinauer, Sunderland.

Vepsalainen, K. (1978). Wing dimorphism and diapause in *Gerris*: Determination and adaptive significance. In *Evolution of insect migration and diapause* (ed. H. Dingle), pp. 218–53. Springer, New York.

Vernberg, F.J., and Vernberg, W.B. (1974). Synergistic effects of temperature and other environmental parameters on organisms. In *Thermal ecology* (eds. J. Gibbons and R. Sharitz), pp. 94–9. US Atomic Energy Commission.

Via, S., and Lande, R. (1987). Evolution of genetic variability in a spatially heterogeneous environment: effects of genotype–environment interaction. *Genetical Research*, **49**, 147–56.

Vrijenhoek, R.L., Douglass, M.E., and Meffe, G.K. (1985). Conservation genetics of endangered fish populations in Arizona. *Science*, **229**, 400–2.

Waddington, C.H. (1953). The genetic assimilation of an acquired character. *Evolution*, **7**, 118–26.

Waddington, C.H. (1956). Genetic assimilation of the bithorax phenotype. *Evolution*, **10**, 1–13.

Waddington, C.H. (1957). *The strategy of the genes*. George Allen and Unwin, London.

Walbot, V., and Cullis, C.A. (1985). Rapid genomic change in higher plants. *Annual Review of Plant Physiology*, **36**, 367–96.

Waldbauer, G.P. (1978). Phenological adaptation and the polymodal emergence patterns of insects. In *Evolution of insect migration and diapause* (ed. H. Dingle), pp. 127–44. Springer-Verlag, New York.

Walker, G.C. (1984). Mutagenesis and inducible responses to deoxyribonucleic acid damage in *Escherichia coli*. *Microbiological Reviews*, **48**, 60–93.

Wallace, A.R. (1876). *The geographical distribution of animals*. MacMillan, London.

Wallace, A.R. (1878). *Tropical nature and other essays*. MacMillan, London.

Wallace, B. (1959). Influence of genetic system on geographical distribution. *Cold Spring Harbour Symposia on Quantitative Biology*, **24**, 193–204.

Wallace, B. (1981). *Basic population genetics*. Columbia University Press, New York.

Wallner, W.E. (1987). Factors affecting insect population dynamics: Differences between outbreak and non-outbreak species. *Annual Review of Entomology*, **32**, 317–40.

Walsh, P.J., and Somero, G.N. (1981). Temperature adaptation in sea anemones: physiological and biochemical variability in geographically separate populations of *Metridium senile*. *Marine Biology*, **62**, 25–34.

Ward, P. (1985). *An analysis of the species' border problem*. PhD Thesis, La Trobe University.

Warwick, E.J. (1972). Genotype–environment interactions in cattle. *World Review of Animal Production*, **8**, 33–8.

Watt, W.B. (1968). The adaptive significance of pigment polymorphism in *Colias* butterflies. *Evolution*, **22**, 437–58.

Watt, W.B. (1983). Adaptation at specific loci. II. Demographic and biochemical elements in the maintenance of *Colias* PGI polymorphism. *Genetics*, **103**, 691–724.

Watt, W.B. (1985). Bioenergetics and evolutionary genetics: Opportunities for a new synthesis. *American Naturalist*, **125**, 118–43.

Wedenmeyer, G.A., McLeay, D.J., and Goodyear, C.P. (1984). Assessing the tolerance of fish and fish populations to environmental stress: the problems and methods of monitoring. *Advances in Environmental Science and Technology*, **16**, 163–95.

Weis, J.S., and Weis, P. (1989). Tolerance and stress in a polluted environment. The case of the mummichog. *BioScience*, **39**, 89–95.

Welden, C.W., and Slauson, W.L. (1986). The intensity of competition versus its importance: an overlooked distinction and some implications. *Quarterly Review of Biology*, **61**, 23–44.

Westerman, J.M., and Parsons, P.A. (1973). Variation in genetic architecture at different doses of γ-radiation as measured by longevity in *Drosophila melanogaster*. *Canadian Journal of Genetics and Cytology*, **15**, 289–98.

White, E.B., Debach, P., and Garber, M.J. (1970). Artificial selection for genetic adaptation to temperature extremes in *Aphytis lingnanensis* (Hymenoptera: Aphelinidae). *Hilgardia*, **40**, 161–92.

White, F.N., and Somero, G. (1982). Acid–base regulation and phospholipid adaptations to temperature: time courses and physiological significance of modifying the milieu for protein function. *Physiological Reviews*, **62**, 40–90.

Wiens, J.A. (1974). Climatic instability and the 'ecological saturation' of bird communities in North American grasslands. *Condor*, **76**, 385–400.

Wiens, J.A. (1977). On competition and variable environments. *American Scientist*, **65**, 590–7.

Wigley, T.M.L., and Atkinson, T.C. (1977). Dry years in south-east England since 1698. *Nature*, **265**, 431–4.

Wijsman, T.C.M. (1976). Adenosine phosphates and energy charge in different tissues of *Mytilus edulis* under aerobic and anaerobic conditions. *Journal of Comparative Physiology*, **107**, 129–40.

Williams, C.K., and Moore, R.J. (1989). Phenotypic adaptation and natural selection in the wild rabbit, *Oryctolagus cuniculus*, in Australia. *Journal of Animal Ecology*, **58**, 495–507.

Williams, G.C. (1966). *Adaptation and natural selection*. Princeton University Press, Princeton.

Wills, C. (1983). The possibility of stress-triggered evolution. *Heredity*, **51**, 530–1.

Wilson, H.R., Wilcox, C.J., Voitle, R.A., Baird, C.D., and Dorminey, R.W. (1975). Characteristics of White Leghorn chickens selected for heat tolerance. *Poultry Science*, **54**, 126–30.

Wilson, J.B. (1988). The cost of heavy-metal tolerance: an example. *Evolution*, **42**, 408–13.

Wilson, S.D., and Keddy, P.A. (1986a). Species competitive ability and position along a natural stress/disturbance gradient. *Ecology*, **67**, 1236–42.

Wilson, S.D., and Keddy, P.A. (1986b). Measuring diffuse competition along an environmental gradient: results from a shoreline plant community. *American Naturalist*, **127**, 862–9.

Woods, K.D., and Davis, M.B. (1989). Paleoecology of range limits: beech in the upper peninsula of Michigan. *Ecology*, **70**, 681–96.

Woodward, F.I. (1987). *Climate and plant distribution*. Cambridge University Press, Cambridge.

Wright, S. (1934). Molecular and evolutionary theories of dominance. *American Naturalist*, **63**, 24–53.

Wright, S., and Dobzhansky, Th. (1946). Genetics of natural populations. XII. Experimental reproduction of some of the changes caused by natural selection in certain populations of *Drosophila pseudoobscura*. *Genetics*, **31**, 125–56.

Yamada, Y., and Bell, A.E. (1969). Selection for larval growth in *Tribolium* under two levels of nutrition. *Genetical Research*, **13**, 175–95.

Yeo, A.R. (1983). Salinity resistance: Physiologies and prices. *Physiologia Plantarum*, **58**, 214–22.

Zachariassen, K.E., Andersen, J., Maloiy, G.M.O. and Kamau, J.M.Z. (1987). Transpiratory water loss and metabolism of beetles from arid areas in East Africa. *Comparative Biochemistry and Physiology*, **86A**, 403–8.

Zhuchenko, A.A., Korol, A.B., and Kovtyukh, L.P. (1985). Change of the crossing-over frequency in *Drosophila* during selection for resistance to temperature fluctuation. *Genetica*, **67**, 73–8.

Zhuchenko, A.A., Korol, A.B., Gavrilenko, T.A., and Kibenko, T.Y. (1986). Correlation between the stability of the genotype and the change in its recombination characteristics under temperature influences. *Genetika*, **22**, 966–74.

Ziolo, L.K. and Parsons, P.A. (1982). Ethanol tolerance, alcohol dehydrogenase activity and *Adh* allozymes in *Drosophila melanogaster*. *Genetica*, **57**, 231–7.

Zivy, M. (1987). Genetic variability for heat shock proteins in common wheat. *Theoretical and Applied Genetics*, **74**, 209–13.

Author index

Subject index